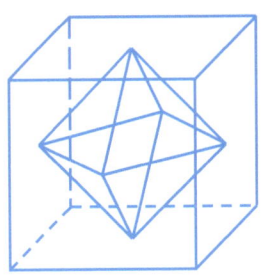 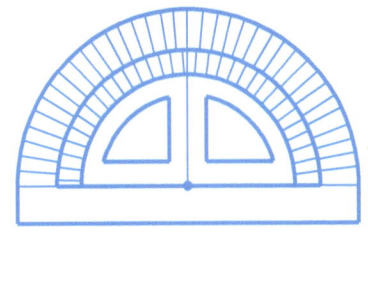 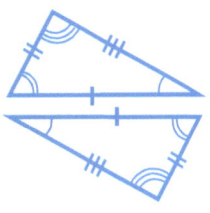 

A Basic Course in Geometry

– Part 3 of 5

Bill Lembke

Citrus Software Publishing

Citrus Ridge, Florida

Published by Citrus Software Publishing, a Division of Citrus Software Corporation

Copyright © 2013 by Citrus Software Corporation

All rights reserved. No part of this publication may be reproduced or distributed in any form or by any means, or stored in a database or retrieval system, without the prior written consent of the publisher, including, but not limited to, network storage or transmission, or broadcast for distance learning.

Printed and bound in the United States of America.

Citrus Software, Citrus Software Publishing, and ABC Method of Instruction are either registered trademarks or trademarks of Citrus Software Corporation in the United Stated and/or other countries. Other products and company names mentioned herein may be the trademarks of their respective owners.

This book expresses the author's views and opinions. The information contained in this book is provided without any express, statutory, or implied warranties. Neither the author, Citrus Software Corporation, nor its resellers, or distributors will be held liable for any damages caused or alleged to be caused either directly or indirectly by this book.

A Basic Course in Geometry – Part 3 of 5

ISBN-13: 978-1477561768

ISBN-10: 1477561765

Table of Contents – Part 3 of 5

Chapter 7 – Polyhedron Solids – Part 1   1

- 7-1 Introduction   1
- 7-2 Platonic Solids   1
- 7-3 Kepler-Poinsot Solids   3
- 7-4 Archimedean Solids   4
- 7-5 Catalan Solids   6
- 7-6 Johnson Solids   8
- 7-7 Summary   12
- Chapter Test   14

Chapter 8 – Polyhedron Solids – Part 2   16

- 8-1 Introduction   16
- 8-2 Pyramids   16
- 8-3 Bipyramids   19
- 8-4 Trapezohedra   21
- 8-5 Frusta   21
- 8-6 Prisms   23
- 8-7 Antiprisms   25
- 8-8 Wedges   25
- 8-9 Toroidal Polyhedra   27
- 8-10 Compound Polyhedra   28
- 8-11 Summary   29
- Chapter Test   31

Chapter 9 – Two Dimensional Non-polytopes   34

- 9-1 Introduction   34
- 9-2 Conic Sections   34
- 9-3 Types of Conics   35
- 9-4 Circles   36
- 9-5 Circular Sectors   39
- 9-6 Circular Segments   39

- 9-7 Circular Rings   40
- 9-8 Circular Ring Sectors   41
- 9-9 Ellipses   42
- 9-10 Parabolas   44
- 9-11 Hyperbolas   45
- 9-12 Stadiums   47
- 9-13 Ovals   47
- 9-14 Summary   48
- Chapter Test   50

Chapter 10- Three Dimensional Non-polytopes   55

- 10-1 Introduction   55
- 10-2 Spheres   55
- 10-3 Spherical Caps   57
- 10-4 Spherical Sectors   57
- 10-5 Spherical Segments   58
- 10-6 Spherical Wedges   59
- 10-7 Cones   59
- 10-8 Cylinders   61
- 10-9 Conical Frusta   61
- 10-10 Tori   62
- 10-11 Capsules   64
- 10-12 Summary   65
- Chapter Test   66

CHAPTER 7: Polyhedron Solids – Part 1

7-1 Introduction

There are many groups or families of polyhedra. This chapter will cover the following groups: Platonic solids, Kepler-Poinsot solids, Archimedean solids, Catalan solids, and Johnson solids.

7-2 Platonic Solids

Platonic solids are polyhedra with regular polygon faces, symmetrical edges, and symmetrical vertices. Each polyhedron is isogonal, isotoxal, isohedral, and convex. These five polyhedra are named after the Greek mathematician Plato. The Greek philosophers and cosmologist of that time period (360 BCE) associated each of the solids with classical elements. The tetrahedron, made up of four equilateral triangles, represented fire. The cube, made up of six squares, represented earth. The octahedron, made up of eight equilateral triangles, represented air. The dodecahedron, made up of twelve pentagons, represented ether or the universe. The icosahedron, made up of twenty equilateral triangles, represented water.

The Platonic solids are the only five convex regular polyhedra. The tetrahedron, cube, octahedron, dodecahedron, and icosahedron are shown in the figures below.

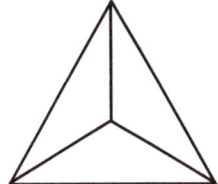
Figure 7-1: Tetrahedron

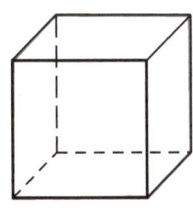
Figure 7-2: Cube

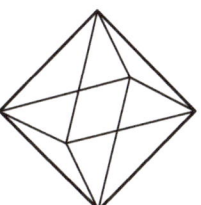
Figure 7-3: Octahedron

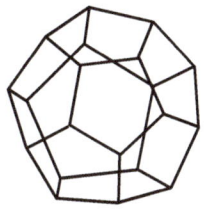
Figure 7-4: Dodecahedron

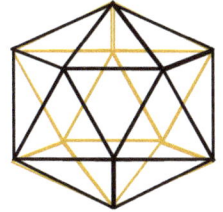
Figure 7-5: Icosahedron

Each of the Platonic solids is a polyhedron with congruent convex regular polygonal faces. As a result of this, there are several important characteristics:

(1) The vertices of the polyhedron all lie on a sphere.

(2) All the dihedral angles are equal.
(3) All the vertex figures are regular polygons.
(4) All the solid angles are equivalent.
(5) All the vertices are surrounded by the same number of faces.

The table below shows formulas of the platonic solids, with (a) representing side length.

Formula	Tetrahedron	Cube	Octahedron	Dodecahedron	Icosahedron
Circumradius	√(3/8)a	(a/2)√3	(a/2)√2	(a/4)(√15+√3)	(a/4)(√10+2(√5))
Inradius	a/√24	a/2	(a/6)√6	(a/20)(√250+110(√5))	(√3)/12(3+√5)a
Midradius	a/√8	(a/2)√2	a/2	(a/4)(3+√5)	(1/4)(1+√5)a
Area	√(3a²)	6a²	2√(3a²)	3(√(25+10(√5)a²))	5(√3)a²
Volume	((√2)/12)a³	a³	(1/3)√(2a³)	(1/4)(15+7(√5)a³	(5/12)(3+√5)a³

Table 7-1: Platonic Solids Formulas – 1

The table below shows values for side length of one unit.

Formula	Tetrahedron	Cube	Octahedron	Dodecahedron	Icosahedron
Circumradius	0.612	0.866	0.707	1.401	0.951
Inradius	0.204	0.500	0.408	1.113	0.755
Midradius	0.353	0.707	0.500	1.309	0.809
Area	1.732	6.000	3.464	20.645	8.660
Volume	0.117	1.000	0.471	7.663	2.181

A dihedral angle is the angle between the faces. A polyhedral angle is the angle between the segments joining the center and the vertices. The table below shows the dihedral and polyhedral angles.

Table 7-2: Platonic Solids Formulas – 2

Angle	Tetrahedron	Cube	Octahedron	Dodecahedron	Icosahedron
Dihedral Angle	70.529	90.000	109.471	116.565	138.190
Polyhedral Angle	109.471	70.529	90.000	41.810	63.435

Table 7-3: Platonic Solids – Polyhedral Angles

7-3 Kepler-Poinsot Solids

Kepler-Poinsot solids are regular polyhedron with regular polygon faces, symmetrical edges, and symmetrical vertices. Each polyhedron is isogonal, isotoxal, isohedral, and concave (stellated). These four polyhedra are named after German mathematician Johannes Kepler and French mathematician Louis Poinsot.

There exist four regular polyhedra which are not convex, called Kepler-Poinsot polyhedra. These all have icosahedral symmetry and may be obtained as stellations of the dodecahedron and the icosahedron. Stellation for polygons and polyhedra is the process of extending edges or faces until they meet to form a new polygon or polyhedron.

The Kepler-Poinsot polyhedra are the small stellated dodecahedron, great stellated dodecahedron, great icosahedron, and great dodecahedron. The first two were discovered by Johannes Kepler and the second two were discovered by Louis Poinsot.

Figure 7-6:
Small Stellated Dodecahedron

Figure 7-7:
Great Stellated Dodecahedron

Figure 7-8:
Great Icosahedron

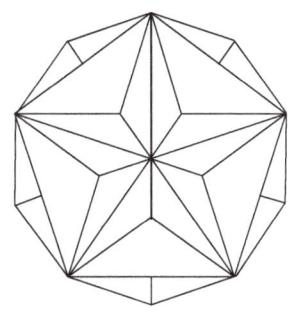
Figure 7-9:
Great Dodecahedron

The table below shows the number of faces, edges, and vertices for each of the four Kepler-Poinsot solids.

Name of Kepler-Poinsot Solid	Faces	Edges	Vertices
Small Stellated Dodecahedron	12	30	12
Great Stellated Dodecahedron	12	30	20
Great Icosahedron	20	30	12
Great Dodecahedron	12	30	12

Table 7-4: Kepler-Poinsot Solids – Faces, Edges, Vertices

7-4 Archimedean Solids

Archimedean solids are convex semiregular polyhedron with regular polygon faces, symmetrical edges, and symmetrical vertices. Unlike a Platonic solid, the faces are made up of two or more different polygons. Each polyhedron is isogonal and convex. These thirteen polyhedra are named after the Greek mathematician Archimedes.

The table below shows the number of faces, edges, and vertices for each of the thirteen Archimedean solids.

Name of Archimedean Solid	Faces	Edges	Vertices	Euler's Formula
Cuboctahedron	14	24	12	12 + 14 = 24 + 2
Icosidodecahedron	32	60	30	30 + 32 = 60 + 2
Truncated Tetrahedron	8	18	12	12 + 8 = 18 + 2
Truncated Octahedron	14	36	24	24 + 14 = 36 + 2
Truncated Cube	14	36	24	24 + 14 = 36 + 2
Truncated Icosahedron	32	90	60	60 + 32 = 90 + 2
Truncated Dodecahedron	32	90	60	60 + 32 = 90 + 2
Rhombicuboctahedron	26	48	24	24 + 26 = 48 + 2
Truncated Cuboctahedron	26	72	48	48 + 26 = 72 + 2
Rhombicosidodecahedron	62	120	60	60 + 62 = 120 + 2
Truncated Icosidodecahedron	62	180	120	120 + 62 = 180 + 2
Snub Cuboctahedron	38	60	24	24 + 38 = 60 + 2
Snub Dodecahedron	92	150	60	60 + 92 = 150 + 2

Table 7-5: Archimedean Solids – Faces, Edges, Vertices

The table below shows the number of each type of regular polygonal face in the solids.

Name of Archimedean Solid	Triangle	Square	Pentagon	Hexagon	Octagon	Decagon
Cuboctahedron	8	6	0	0	0	0
Icosidodecahedron	20	0	12	0	0	0
Truncated Tetrahedron	4	0	0	4	0	0
Truncated Octahedron	0	6	0	8	0	0
Truncated Cube	8	0	0	0	6	0
Truncated Icosahedron	0	0	12	20	0	0
Truncated Dodecahedron	20	0	0	0	0	12
Rhombicuboctahedron	8	18	0	0	0	0

Truncated Cuboctahedron	0	12	0	8	6	0
Rhombicosidodecahedron	20	30	12	0	0	0
Truncated Icosidodecahedron	0	30	0	20	0	12
Snub Cuboctahedron	32	6	0	0	0	0
Snub Dodecahedron	80	0	12	0	0	0

Table 7-6: Archimedean Solids – Number of Faces by Type

All Archimedean solids can be thought of as modifications of Platonic solids. The processes of truncation, rectification, snubification, and expansion can be applied to get the desired result.

- Truncation: Two points are cut on an edge at each vertex, such that the vertex is cut off. For example a cube becomes a truncated cube with 14 regular faces, 6 of which are octagons and the other 8 are equilateral triangles.
- Rectification: A form of truncation with one point cut on an edge at each vertex, such that the vertex is cut off. For example, a cube becomes a cuboctahedron with 14 regular faces, 6 of which are squares and the other 8 are equilateral triangles.
- Snubification: A form of expansion of a polyhedron in which all of the faces are slightly rotated in the same direction. The empty spaces are filled with triangles. For example, a cube becomes a snub cuboctahedron with 38 regular faces, 6 of which are squares and the other 32 are equilateral triangles.
- Expansion: The faces are moved outward from the center of a polyhedron, with the vertices connected across the empty spaces. For example, a cube becomes a rhombicubocahedron with 26 regular faces, 18 of which are squares and the other 8 are equilateral triangles.

The figures below show a cube transformed into four of the Archimedean solids.

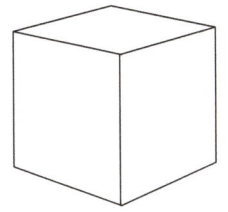
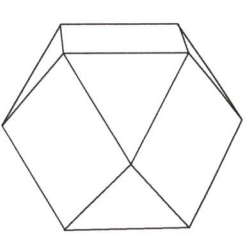
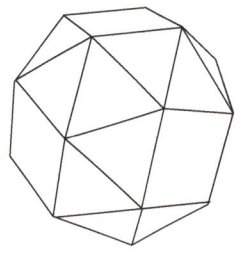
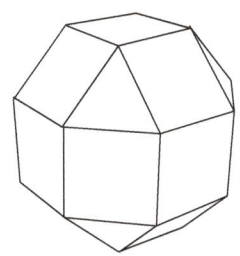

Figure 7-10: Cube

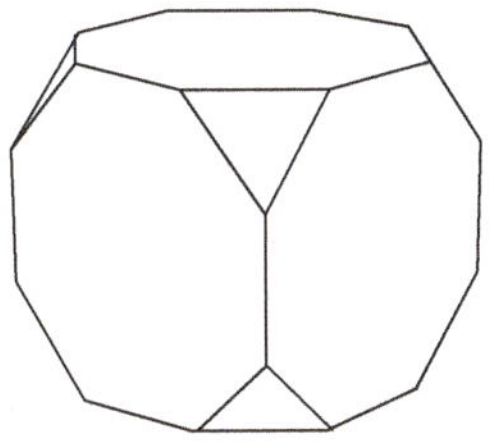

Figure 7-11: Truncation – Truncated Cube

Figure 7-12: Rectification – Cuboctahedron

Figure 7-13: Snubification – Snub Cuboctahedron

Figure 7-14: Expansion – Rhombicubocahedron

The Archimedean solids can be grouped by common characteristics as follows.

- Without triangular faces (4): truncated icosahedron, truncated octahedron, rhombicosidodecahedron, and truncated icosidodecahedron
- Three edges meet at each vertex (7): truncated icosahedron, truncated octahedron, rhombicosidodecahedron, truncated icosidodecahedron, truncated tetrahedron, truncated cube, and truncated dodecahedron
- Truncated Platonic solids (5): truncated icosahedron, truncated octahedron, truncated tetrahedron, truncated cube, and truncated dodecahedron
- Four edges meet at each vertex (4): cuboctahedron, rhombicuboctahedron, icosidodecahedron, and rhombicosidodecahedron
- Five edges meet at each vertex (2): snub cuboctahedron and snub dodecahedron
- Chiral – mirror images or two versions (2): snub cuboctahedron and snub dodecahedron

The truncated icosahedron is the shape of a traditional soccer ball.

7-5 Catalan Solids

Catalan solids are dual polyhedrons to the Archimedean solids. The Catalan solids are convex and isohedral. Unlike the Platonic solids and the Archimedean solids, the faces are not regular polygons. Each face is identical. These thirteen polyhedra are named after Belgian mathematician Eugene Catalan.

The table below shows the number of faces, edges, and vertices for each of the thirteen Catalan solids.

Name of Catalan Solid	Faces	Edges	Vertices	Euler's Formula
Triakis Tetrahedron	12	18	8	8 + 12 = 18 + 2
Rhombic Dodecahedron	12	24	14	14 + 12 = 24 + 2
Triakis Octahedron	24	36	14	14 + 24 = 36 + 2
Tetrakis Hexahedron	24	36	14	14 + 24 = 36 + 2
Deltoidal Icositetrahedron	24	48	26	26 + 24 = 48 + 2
Disdyakis Dodecahedron	48	72	26	26 + 48 = 72 + 2
Pentagonal Icositetrahedron	24	60	38	38 + 34 = 60 + 2
Rhombic Triacontahedron	30	60	32	32 + 30 = 60 + 2
Triakis Icosahedron	60	90	32	32 + 60 = 90 + 2
Pentakis Dodecahedron	60	90	32	32 + 60 = 90 + 2
Deltoidal Hexecontahedron	60	120	62	62 + 60 = 120 + 2
Disdyakis Triacontahedron	120	180	62	62 + 120 = 180 + 2
Pentagonal Hexecontahedron	60	150	92	92 + 60 = 150 + 2

Table 7-7: Catalan Solids– Faces, Edges, Vertices

The table below shows describes the polygonal faces in the solids.

Name of Catalan Solid	Faces	Face Description
Triakis Tetrahedron	12	120-30-30 triangle
Rhombic Dodecahedron	12	120-60-120-60 rhombus
Triakis Octahedron	24	120-30-30 triangle
Tetrakis Hexahedron	24	90-45-45 triangle
Deltoidal Icositetrahedron	24	1:1.29 side ratio kite
Disdyakis Dodecahedron	48	90-54-36 triangle
Pentagonal Icositetrahedron	24	(2 sides) 0.842:0.593 (3 sides) side ratio pentagon
Rhombic Triacontahedron	30	116.57-63.43-116.57-63.43 rhombus
Triakis Icosahedron	60	120-30-30 triangle
Pentakis Dodecahedron	60	72-54-54 triangle
Deltoidal Hexecontahedron	60	1:1.54 side ratio kite
Disdyakis Triacontahedron	120	90-54-36 triangle
Pentagonal Hexecontahedron	60	(2 sides) 1.019:0.582 (3 sides) side ratio pentagon

Table 7-8: Catalan Solids – Number of Faces by Type

The figures below show examples of Catalan solids.

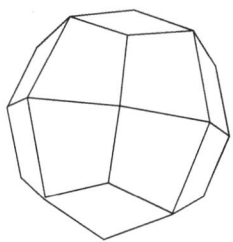 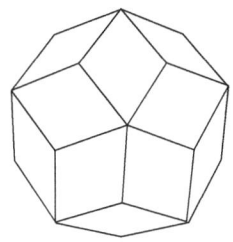 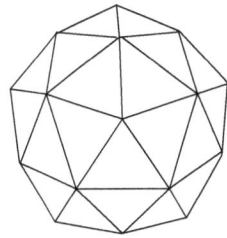 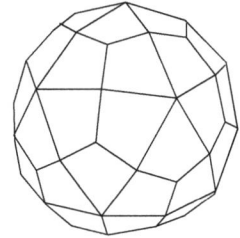

Figure 7-15: Deltoidal Icositetrahedron

Figure 7-16: Rhombic Triacontahedron

Figure 7-17: Pentakis Dodecahedron

Figure 7-18: Deltiodal Hexecontahedron

7-6 Johnson Solids

Johnson solids are convex polyhedron with regular polygon faces and equal edge lengths, but are not isogonal. Unlike a Platonic solid, the faces are made up of two or more different polygons. These ninety-two polyhedra are named after the American mathematician Norman Johnson.

A systematic nomenclature or naming system for the polyhedra can be used to describe the characteristics, such as the shape of the sides and how they are connected to each other. A list of descriptive terms or adjectives follows.

- **Bi- or Di-**: This prefix means two copies of a solid are joined base to base. Cupolae and rotundae can be joined either with like faces meeting (ortho-) or unlike faces meeting (gyro-). For example, an octahedron is a square bipyramid, a cuboctahedron is a triangular gyrobicupula, and an icosadodecahedron is a pentagonal gyrobirotunda.
- **Elongated**: A prism is joined to the base of a solid or between the bases of a solid. For example, a rhombicuboctahedron is an elongated square orthobicupola.
- **Gyroelongated**: An antiprism is joined to the base of a solid or between the bases of a solid. For example, an icosahedron is a gyroelongated pentagonal bipyramid.
- **Augmented**: A pyramid or cupola is joined to a face of a solid.
- **Diminished**: A pyramid or cupola is removed from a solid.
- **Gyrate**: A cupola on a solid is rotated so different edges match up, as in the difference between orthobicupola and gyrobicupola.
- **Sphenoid**: A wedge or half prism that includes the base polygon and its featured edge.
- **Hemiprism**: A half prism where the cutting plane contains only a single vertex, instead of a whole edge.
- **Para-**: The augmented faces are parallel.
- **Meta-**: The solid has two oblique faces augmented. Oblique faces are not right angles or a multiple of right angle.

The figures below show examples of Johnson solids.

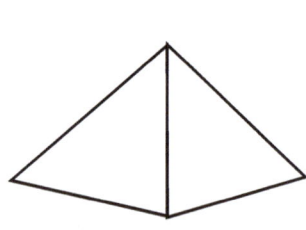

Figure 7-19:
Square Pyramid

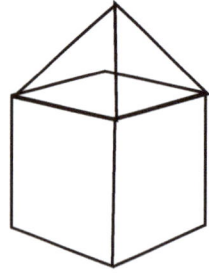

Figure 7-20:
Elongated
Square Pyramid

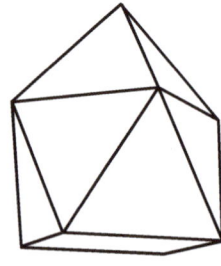

Figure 7-21:
Gyroelongated
Square Pyramid

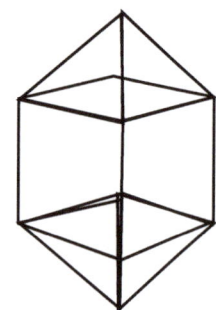

Figure 7-22:
Elongated Square
Dipyramid

The table below shows the number of each type of regular polygonal face in the solids.

Name of Johnson Solid	Triangle	Square	Pentagon	Hexagon	Octagon	Decagon
Square pyramid	4	1	0	0	0	0
Pentagonal pyramid	5	0	1	0	0	0
Triangular cupola	4	3	0	1	0	0
Square cupola	4	5	0	0	1	0
Pentagonal cupola	5	5	1	0	0	1
Pentagonal rotunda	10	0	6	0	0	1
Elongated triangular pyramid	4	3	0	0	0	0
Elongated square pyramid	4	5	0	0	0	0
Elongated pentagonal pyramid	5	5	1	0	0	0
Gyroelongated square pyramid	12	1	0	0	0	0
Gyroelongated pentagonal pyramid	15	0	1	0	0	0
Triangular dipyramid	6	0	0	0	0	0
Pentagonal dipyramid	10	0	0	0	0	0
Elongated triangular dipyramid	6	3	0	0	0	0
Elongated square dipyramid	8	4	0	0	0	0
Elongated pentagonal dipyramid	10	5	0	0	0	0
Gyroelongated square dipyramid	16	0	0	0	0	0
Elongated triangular cupola	4	9	0	1	0	0
Elongated square cupola	4	13	0	0	1	0
Elongated pentagonal cupola	5	15	1	0	0	1
Elongated pentagonal rotunda	10	10	6	0	0	1
Gyroelongated triangular cupola	16	3	0	1	0	0
Gyroelongated square cupola	20	5	0	0	1	0
Gyroelongated pentagonal cupola	25	5	1	0	0	1
Gyroelongated pentagonal rotunda	30	0	6	0	0	1

Gyrobifastigium	4	4	0	0	0	0
Triangular orthobicupola	8	6	0	0	0	0
Square orthobicupola	8	10	0	0	0	0
Square gyrobicupola	8	10	0	0	0	0
Pentagonal orthobicupola	10	10	2	0	0	0
Pentagonal gyrobicupola	10	10	2	0	0	0
Pentagonal orthocupolarotunda	15	5	7	0	0	0
Pentagonal gyrocupolarotunda	15	5	7	0	0	0
Pentagonal orthobirotunda	20	0	12	0	0	0
Elongated triangular orthobicupola	8	12	0	0	0	0
Elongated triangular gyrobicupola	8	12	0	0	0	0
Elongated square gyrobicupola	8	18	0	0	0	0
Elongated pentagonal orthobicupola	10	20	2	0	0	0
Elongated pentagonal gyrobicupola	10	20	2	0	0	0
Elongated pentagonal orthocupolarotunda	15	15	7	0	0	0
Elongated pentagonal gyrocupolarotunda	15	15	7	0	0	0
Elongated pentagonal orthobirotunda	20	10	12	0	0	0
Elongated pentagonal gyrobirotunda	20	10	12	0	0	0
Gyroelongated triangular bicupola	20	6	0	0	0	0
Gyroelongated square bicupola	24	10	0	0	0	0
Gyroelongated pentagonal bicupola	30	10	2	0	0	0
Gyroelongated pentagonal cupolarotunda	35	5	7	0	0	0
Gyroelongated pentagonal birotunda	40	0	12	0	0	0
Augmented triangular prism	6	2	0	0	0	0
Biaugmented triangular prism	10	1	0	0	0	0
Triaugmented triangular	14	0	0	0	0	0

prism						
Augmented pentagonal prism	4	4	2	0	0	0
Biaugmented pentagonal prism	8	3	2	0	0	0
Augmented hexagonal prism	4	5	0	2	0	0
Parabiaugmented hexagonal prism	8	4	0	2	0	0
Metabiaugmented hexagonal prism	8	4	0	2	0	0
Triaugmented hexagonal prism	12	3	0	2	0	0
Augmented dodecahedron	5	0	11	0	0	0
Parabiaugmented dodecahedron	10	0	10	0	0	0
Metabiaugmented dodecahedron	10	0	10	0	0	0
Triaugmented dodecahedron	15	0	9	0	0	0
Metabidiminished icosahedron	10	0	2	0	0	0
Tridiminished icosahedron	5	0	3	0	0	0
Augmented tridiminished icosahedron	7	0	3	0	0	0
Augmented truncated tetrahedron	8	3	0	3	0	0
Augmented truncated cube	12	5	0	0	5	0
Biaugmented truncated cube	16	10	0	0	4	0
Augmented truncated dodecahedron	25	5	1	0	0	11
Parabiaugmented truncated dodecahedron	30	10	2	0	0	10
Metabiaugmented truncated dodecahedron	30	10	2	0	0	10
Triaugmented truncated dodecahedron	35	15	3	0	0	9
Gyrate rhombicosidodecahedron	20	30	12	0	0	0
Parabigyrate rhombicosidodecahedron	20	30	12	0	0	0
Metabigyrate rhombicosidodecahedron	20	30	12	0	0	0
Trigyrate rhombicosidodecahedron	20	30	12	0	0	0

Diminished rhombicosidodecahedron	15	25	11	0	0	1
Paragyrate diminished rhombicosidodecahedron	15	25	11	0	0	1
Metagyrate diminished rhombicosidodecahedron	15	25	11	0	0	1
Bigyrate diminished rhombicosidodecahedron	15	25	11	0	0	1
Parabidiminished rhombicosidodecahedron	10	20	10	0	0	2
Metabidiminished rhombicosidodecahedron	10	20	10	0	0	2
Gyrate bidiminished rhombicosidodecahedron	10	20	10	0	0	2
Tridiminished rhombicosidodecahedron	5	15	9	0	0	3
Snub disphenoid	12	0	0	0	0	0
Snub square antiprism	24	2	0	0	0	0
Sphenocorona	12	2	0	0	0	0
Augmented sphenocorona	16	1	0	0	0	0
Sphenomegacorona	16	2	0	0	0	0
Hebesphenomegacorona	18	3	0	0	0	0
Disphenocingulum	20	4	0	0	0	0
Bilunabirotunda	8	2	4	0	0	0
Triangular hebesphenorotunda	13	3	3	1	0	0

Table 7-9: Johnson Solids – Number of Faces by Type

• 7-7 Summary

Platonic solids are polyhedra with regular polygon faces, symmetrical edges, and symmetrical vertices. Each polyhedron is isogonal, isotoxal, isohedral, and convex. The Platonic solids are the only five convex regular polyhedra. These five polyhedra are the tetrahedron, cube, octahedron, dodecahedron, and icosahedron. Each of the Platonic solids is a polyhedron with congruent convex regular polygonal faces.

Kepler-Poinsot solids are regular polyhedron with regular polygon faces, symmetrical edges, and symmetrical vertices. Each polyhedron is isogonal, isotoxal, isohedral, and concave (stellated). There exist four regular polyhedra which are not convex, called Kepler-Poinsot polyhedra. These all have icosahedral symmetry and may be obtained as stellations of the dodecahedron and the icosahedron. Stellation for polygons and polyhedra is the process of extending edges or faces until they meet to form a new polygon or polyhedron.

Archimedean solids are convex semiregular polyhedron with regular polygon faces, symmetrical edges, and symmetrical vertices. Unlike a Platonic solid, the faces are made up of two or more different polygons. Each polyhedron is isogonal and convex. All Archimedean solids can be thought of as modifications of Platonic solids. The processes of truncation, rectification, snubification, and expansion can be applied to get the desired result.

Catalan solids are dual polyhedrons to the Archimedean solids. The Catalan solids are convex and isohedral. Unlike the Platonic solids and the Archimedean solids, the faces are not regular polygons. Each face is identical.

Johnson solids are convex polyhedron with regular polygon faces and equal edge lengths, but are not isogonal. Unlike a Platonic solid, the faces are made up of two or more different polygons.

CHAPTER 7

Chapter Test

Grading Scale: One point for each correct answer.

Excellent = 46-51, Good = 41-45, Average = 36-40, Fair = 31-35, Poor = 0-30

• 7-2 Platonic Solids

Calculate circumradius, inradius, and midradius. (Round to the nearest thousandth.)

1. Tetrahedron with side = 6 _____ _____ _____
2. Cube with side = 5 _____ _____ _____
3. Octahedron with side = 4 _____ _____ _____
4. Dodecahedron with side = 3 _____ _____ _____
5. Icosahedron with side = 2 _____ _____ _____

Calculate surface area and volume. (Round to the nearest thousandth.)

6. Tetrahedron with side = 6 _____ _____
7. Cube with side = 5 _____ _____
8. Octahedron with side = 4 _____ _____
9. Dodecahedron with side = 3 _____ _____
10. Icosahedron with side = 2 _____ _____

• 7-3 Kepler-Poinsot Solids

Mark as True or False.

1. Each polyhedron is isogonal, isotoxal, isohedral. _____
2. Kepler-Poinsot solids are regular polyhedra which are convex. _____
3. These six polyhedra are named after Italian mathematician Kepler Poinsot. _____
4. All Kepler-Poinsot solids have 30 edges. _____

• 7-4 Archimedean Solids

Match definitions and terms.

A = Truncation   B = Rectification   C = Snubification
D = Expansion   E = Chiral

1. Mirror images or two versions. ____
2. Faces are moved outward, with the vertices connected across the empty spaces. ____
3. Two points are cut on an edge at each vertex, such that the vertex is cut off. ____
4. One point cut on an edge at each vertex, such that the vertex is cut off. ____
5. Faces are moved outward, vertices connected, and all faces rotated. ____

7-5 Catalan Solids

Mark as True or False.

1. Catalan solids are dual polyhedrons to the Archimedean solids. ____
2. The Catalan solids are convex and isohedral, with regular polygon faces. ____
3. Each face is identical. ____
4. All faces must be a triangle, rhombus, kite, or pentagon. ____

7-6 Johnson Solids

Match definitions and terms.

A = Bi or Di   B = Ortho   C = Gyro   D = Elongated
E = Gyroelongated   F = Augmented   G = Diminished   H = Gyrate
I = Sphenoid   J = Hemiprism   K = Para   L = Meta   M = Oblique

1. The augmented faces are parallel. ____
2. A pyramid or cupola is joined to a face of a solid. ____
3. A cupola on a solid is rotated so different edges match up. ____
4. Two copies of a solid are joined base to base. ____
5. Unlike faces meeting. ____
6. Faces are not right angles or a multiple of right angle. ____
7. An antiprism is joined to the base of a solid or between the bases of a solid. ____
8. A half prism where the plane contains single vertex, instead of a whole edge. ____
9. Like faces meeting. ____
10. The solid has two oblique faces augmented. ____
11. A pyramid or cupola is removed from a solid. ____
12. A prism is joined to the base of a solid or between the bases of a solid. ____
13. A wedge or half prism that includes the base polygon and its featured edge. ____

CHAPTER 8
Polyhedron Solids – Part 2

• 8-1 Introduction

There are many groups or families of polyhedra. This chapter will cover the following groups: pyramids, bipyramids, trapezohedron, frustum, prism, antiprism, wedge, toroidal polyhedron, and compound polyhedron.

• 8-2 Pyramids

A pyramid is a polyhedron in the form of a conic solid. A conic solid is a three-dimensional geometric shape bounded by a plane base and the surface formed by line segments connecting the perimeter of the base to a common point, or apex, outside the plane of the base. A conic solid with a polygon base is a pyramid. A conic solid with a circular base is a cone. The figures below show conic sections.

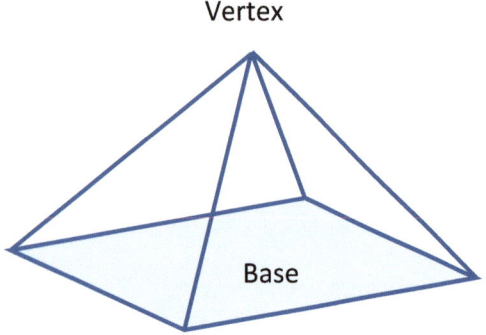

Figure 8-1: Conic Solid – Pyramid

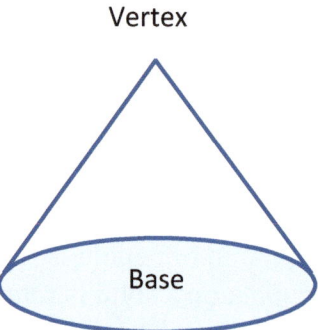

Figure 8-2: Conic Solid – Cone

The base of a pyramid can be any polygon shape. The sides of the pyramid, called lateral faces, are always triangular shaped. The segments where the lateral faces meet are called lateral edges. The apex where all of the sides meet is called the vertex of the pyramid. The height of a pyramid is the perpendicular distance from the vertex to the base. The figures below show the parts of a pyramid.

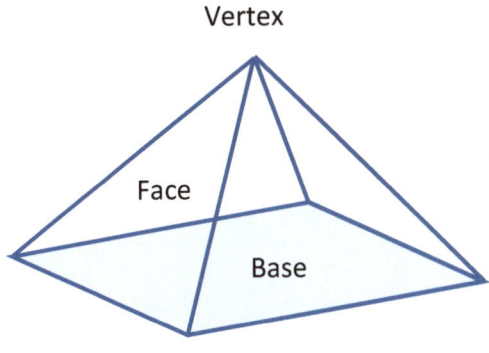
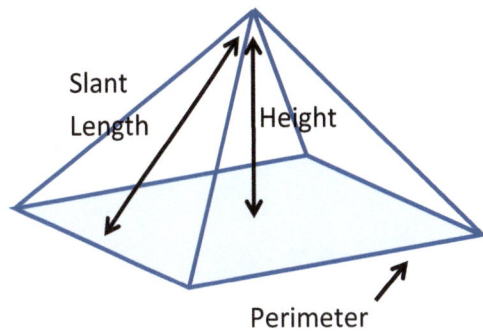

Figure 8-3: Pyramid – Parts   Figure 8-4: Pyramid – Measurements

Pyramids can be classified based on the location of the vertex and by the shape of the base polygon. Pyramids can be right or oblique. If an axis through the vertex and the center of the base, the altitude, intersects the base perpendicularly, then the pyramid is a right pyramid. If the axis does not intersect perpendicularly, then the pyramid is an oblique pyramid. In a right pyramid, the height and altitude are the same line or axis. The figures below show right and oblique pyramids.

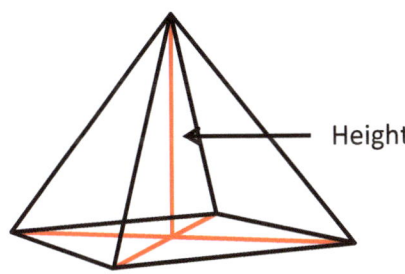
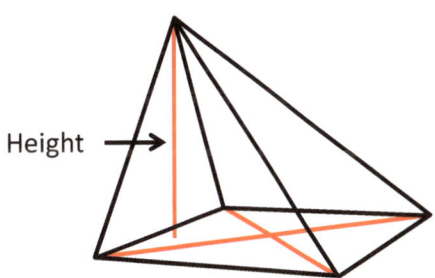

Figure 8-5: Right Pyramid   Figure 8-6: Oblique Pyramid

The shape of the base polygon of a pyramid can be regular or irregular. The base of a regular pyramid is a regular equilateral and equiangular polygon. The lateral faces of a regular pyramid are congruent isosceles triangles. The triangular pyramid (tetrahedron), square pyramid, and pentagonal pyramid are the only regular pyramids whose faces can be equilateral triangles. The base of an irregular pyramid is an irregular polygon. Pyramids with regular star polygon bases are called star pyramids. For example, the pentagram star pyramid has a

17

pentagram base and five intersecting triangle sides. The figures below show regular based pyramid, irregular based pyramid, and star based pyramids.

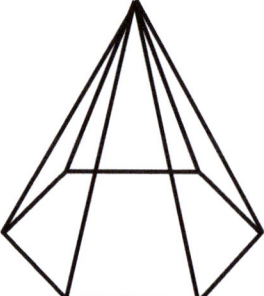

Figure 8-7:
Hexagonal Pyramid

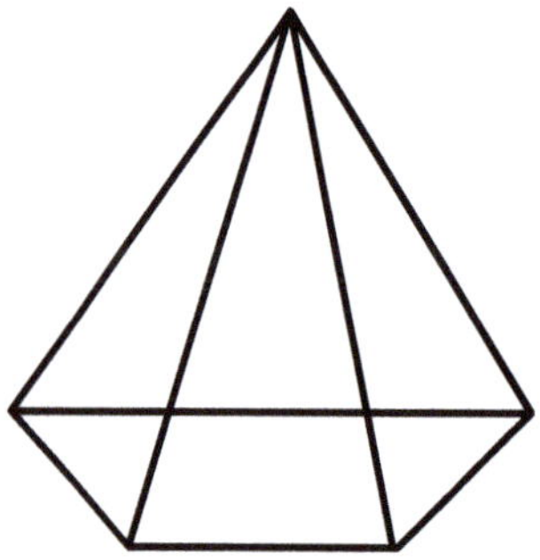

Figure 8-8:
Irregular Pyramid

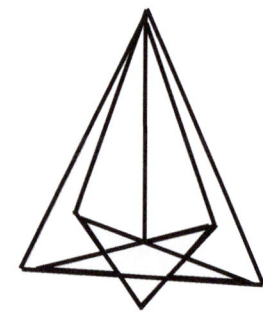

Figure 8-9:
Pentagram Star Pyramid

The surface area and volume of a pyramid can be calculated.

The center of mass of a pyramid is at a height of 1/4 of the total height, from the base. A pyramid with an n-sided base will have $n + 1$ vertices, $n + 1$ faces, and $2n$ edges. All pyramids are self-dual. When unspecified, the base is usually assumed to be square. The first figure below shows a square pyramid with surface area and volume formulas. The second figure below shows a square pyramid with sample values.

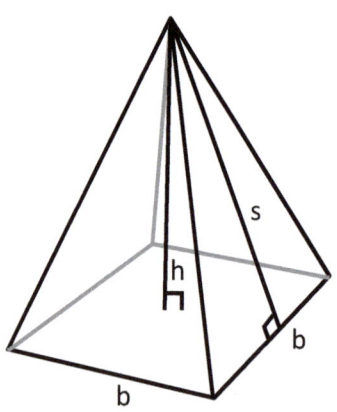

Surface Area: $A = 2bs + b^2$

Volume: $1/3\ b^2 h$

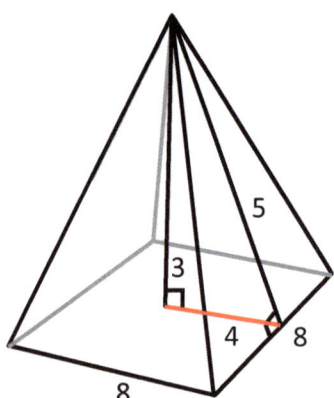

Figure 8-10: Square Pyramid with Formulas

Figure 8-11: Sample Square Pyramid

The surface area of a pyramid is the sum the base area and each of the face areas. If the side faces are the same, as in a regular pyramid, the formula can be written as: base area + 1/2 x perimeter x slant length. If the side faces are different, as in an irregular pyramid, the formula can be written as: base area + lateral area, with the lateral area being the sum of each face area.

The following example calculates the surface area for a square base pyramid. If the length of the base is 8 units, then the measurement from the center of the base to the side of the base is 4 units. The height of the pyramid is 3 units. Since the altitude intersects the base at a perpendicular angle, the slant length can be determined by using the Pythagorean Theorem. The formula can be written as: $3^2 + 4^2 = 5^2$, which means the slant length is 5 units. Using the first formula, base area + 1/2 x perimeter x slant length, the surface area is calculated as: (8 x 8) + 1/2 x 32 x 5. Using the second formula, base area + lateral area, the surface area is calculated as: (8 x 8) + 4 x (1/2 x 5 x 4) x 2). The result of both formulas is a surface area of 144 square units.

The volume of a pyramid is equal to one third the product of the area of the base and the altitude and can be written as: 1/3 x base area x height. Since the base area is the length of the base times the width of the base, this formula can also be expressed as 1/3 x L x W x H.

The following example calculates the volume for a square base pyramid. If the length of the base is 8 units and the height of the pyramid is 3 units, then volume can be calculated as: 1/3 x 64 x 3 or 1/3 x 8 x 8 x 3. The result of both formulas is a volume of 64 cubic units.

The volume of a regular square pyramid, which has equilateral triangle sides, can be calculated differently. The height of a regular square pyramid is 1/2 the base length. The formula can be written as: h = 1/2 s, with s for side length. This means its volume is 1/6 the volume of a cube with the same base dimensions. Since the formula for the area of a cube is s^3, the formula for the volume of a regular square pyramid can then be expressed as $s^3/6$.

The following example calculates the volume for a regular square pyramid. If the length of the base is 8 units, then the height is 4 units. Using the first formula, which applies to all regular base pyramids, 1/3 x L x W x H, the volume can be calculated as: 1/3 x 8 x 8 x 4. Using the second formula, which applies to regular square pyramids with equilateral triangle sides, $s^3/6$, the volume can be calculated as: $8^3 / 6$. The result of both formulas is a volume of 85.3333 cubic units.

8-3 Bipyramids

A bipyramid or dipyramid is a polyhedron created by attaching two pyramids symmetrically base to base. The resulting shape has two apices. A line, such as the altitude, passing through the two apices will intersect the base plane at the center of the bases and at a perpendicular angle. The distance from each apex to the base is equal. The figures below show two square pyramids and a square bipyramid. A regular square bipyramid is also called an octahedron.

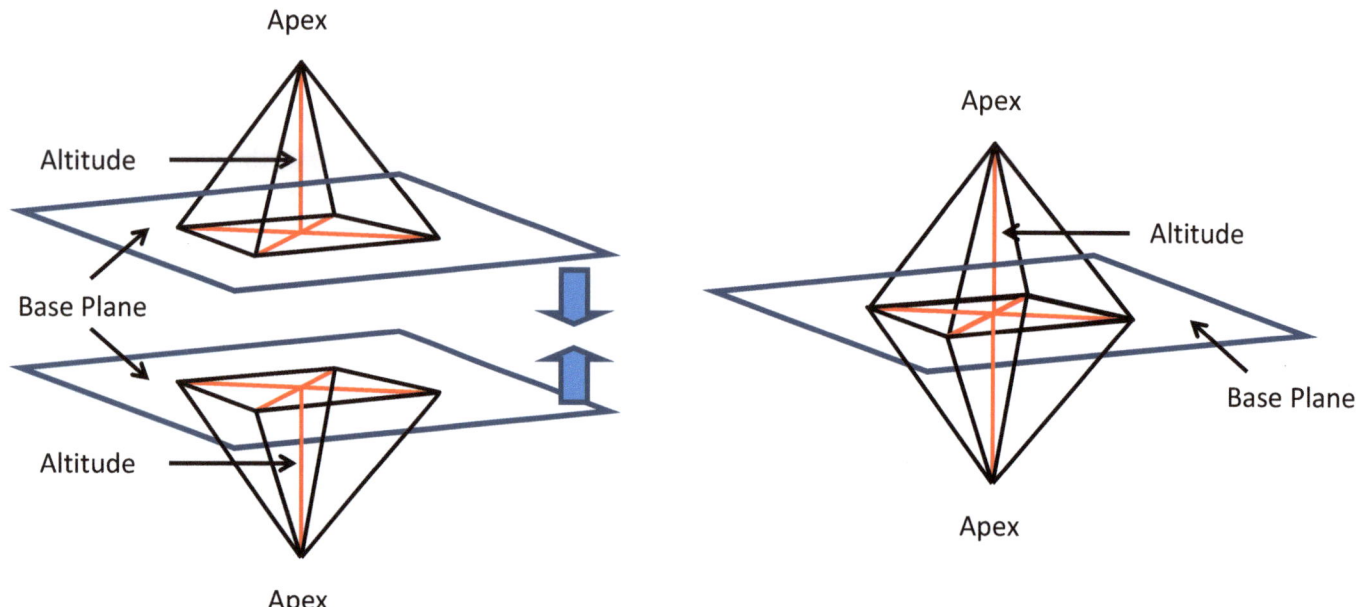

Figure 8-12: Two Square Pyramids   Figure 8-13: Square Bipyramid

The surface area of a bipyramid is twice that of two pyramids, minus two times the base area. The surface area is the sum of all the lateral face areas.

The volume of a bipyramid is equal to two thirds the product of the area of the base and the height and can be written as: 2/3 x base area x height, or $2/3\ b^2h$. The height is the distance from the base to an apex. Since a bipyramid is two pyramids attached together, the volume of a bipyramid is twice that of a pyramid.

The triangular bipyramid, square bipyramid, and pentagonal bipyramid are the only regular bipyramids whose faces can be equilateral triangles. This is because they are made correspondingly from triangular pyramids, square pyramids, and pentagonal pyramids. Star bipyramids are two star pyramids attached symmetrically base to base.

The center of mass of a bipyramid is at the center of the base. A bipyramid with an n-sided base will have $n + 2$ vertices, $2n$ faces, and $3n$ edges. All bipyramids are self-dual.

8-4 Trapezohedra

A <mark>trapezohedron</mark> or <mark>antidipryramid</mark> is a polyhedron with faces composed of congruent kites. A trapazohedron is similar to a bipyramid, but instead of triangular sides, it has kite sides. However, instead of attaching the two halves symmetrically like in a bipyramid, the two halves of a trapezohedron are slightly off set to allow the vertices to match up. Trapezonedra are the dual polyhedra of the Archimedean antiprisms.

The n-gon part of the name does not reference the faces here but arrangement of vertices around an axis of symmetry, or the number of faces in one half of the trapezohedron. The <mark>center of mass</mark> of a trapezohedron is at the center of the base. A trapezohedron with an n-vertice base will have $2n + 2$ vertices, $2n$ faces, and $4n$ edges. An n-gonal trapezohedron can be decomposed into two equal n-gonal pyramids and an n-gonal antiprism. A trapezohedron is convex and isohedral or face transitive. The figures below show a tetragonal trapezohedron, pentagonal trapezohedron, hexagonal trapezohedron, and decagonal trapezohedron.

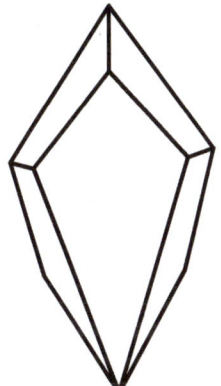
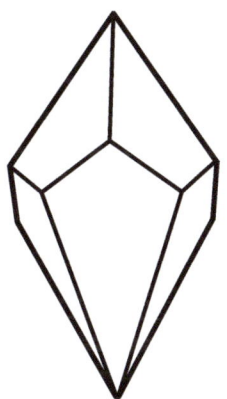
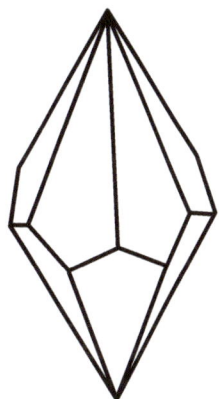
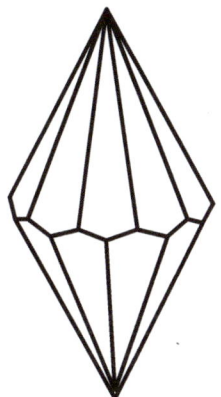

Figure 8-14: Tetragonal Trapezohedron

Figure 8-15: Pentagonal Trapezohedron

Figure 8-16: Hexagonal Trapezohedron

Figure 8-17: Decagonal Trapezohedron

8-5 Frusta

A <mark>frustum</mark> is a polyhedron created by cutting off the top of a conic solid with a plane parallel to the base of the solid. The portion between the top plane and the base plane is the frustum. If the original pyramid was a right pyramid, the associated frustum is a right frustum. An axis through the center of the upper base and the lower base intersects the bottom base perpendicularly. If the original pyramid was oblique, the associated frustum is oblique. The axis does not intersect the base perpendicularly. The figures below show frustum for conic solids.

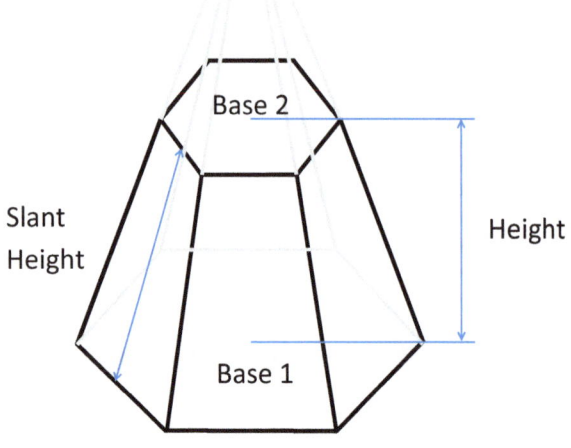

Figure 8-18: Pyramid Frustum

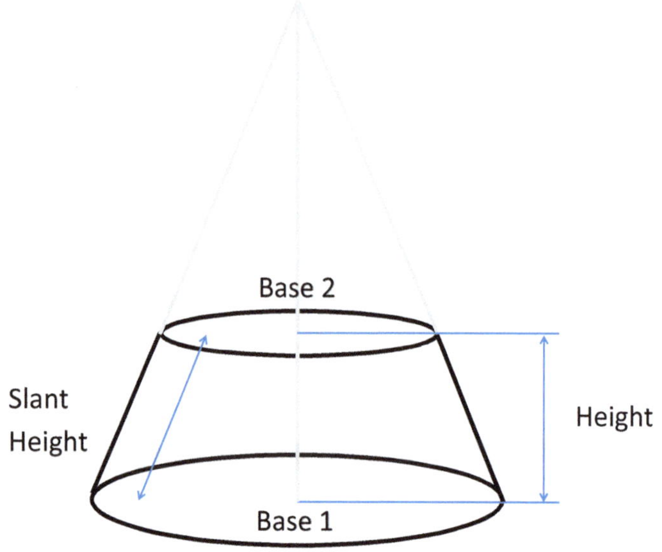

Figure 8-19: Cone Frustum

The height of a frustum is the perpendicular distance between the planes of the two bases. Two frustra joined together make a bifrustum. A frustum with n-gons as bases will have n-gon +2 faces, 3n edges, and 2n vertices. The surface area of a right pyramidal frustum is one half of the sum of the perimeters of the bases times the slant height and can be written as: 1/2 (p1 + p2) s. P1 is the perimeter of base 1, the lower base. P2 is the perimeter of base 2, the upper base. S is the slant height. The surface area of an oblique pyramidal frustum is sum of the lateral areas of the two bases and the face lateral areas. The volume of pyramidal frustum is equal to the one-third product of the altitude and the sum of the upper base area, the lower base area, and the square root of the product of the two base areas. This can be written as: 1/3 h (A1 + A2 + √(A1A2)). H is the height or altitude. A1 is area of the lower or bottom base. A2 is the area of the upper or top base. The first figure below shows a right pyramidal frustum with labeled areas. The second figure below shows a right square pyramidal frustum with sample values.

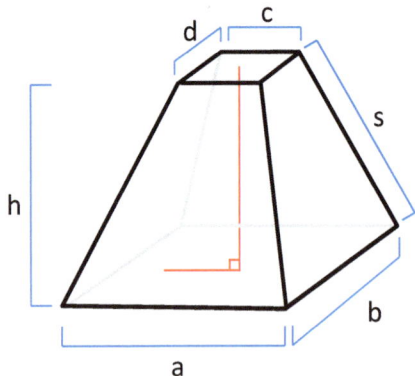

Surface Area: A = 1/2 (p1 + p2) s

Volume: 1/3 h x ((a x b) + (c x d) + √((a x b) (c x d)))

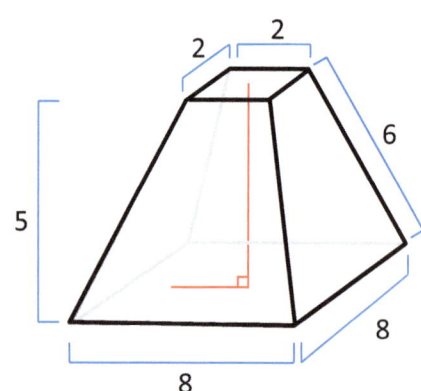

Figure 8-20:
Right Square Pyramidal
Frustum with Formulas

Figure 8-21:
Sample Right Square
Pyramidal Frustum

The following example calculates the surface area for a right pyramidal frustum. The bottom base measures 8 by 8 units and the top base measures 2 by 2 units. The height is 5 units and the slant height is 6 units. The surface area formula can be written as: 1/2 (p1 + p2) s. The surface area is calculated as: 1/2 (32 + 8) 6. The result of the formula is 120 square units.

The following example calculates the volume of for a right pyramidal frustum. The volume formula can be written as: 1/3 h (A1 + A2 + √(A1A2)). Since the pyramid is square, the area of each base is the length times the width. The volume is calculated as: 1/3 x 5 (64 + 4 + √(64 x 4)). The result of the formula is 140 cubic units.

8-6 Prisms

A prism is a polyhedron with two identical *n*-sided polygonal bases and *n* other parallelogram lateral faces connecting the corresponding sides of the bases. The two bases have the same size and shape. All cross-sections of the prism parallel to a base are the same. Prisms are named according to the shape of their base or cross-section, so a prism with a triangular base is called a triangular prism.

A right prism is a prism which has one base aligned directly above the other base. The lateral faces are perpendicular to the bases and rectangular. If the lateral faces are not perpendicular to the bases, then the prism is an oblique prism. The lateral faces are parallelograms, but not rectangles. The figures below show a right prism and an oblique prism.

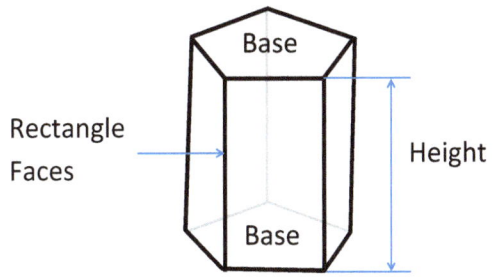

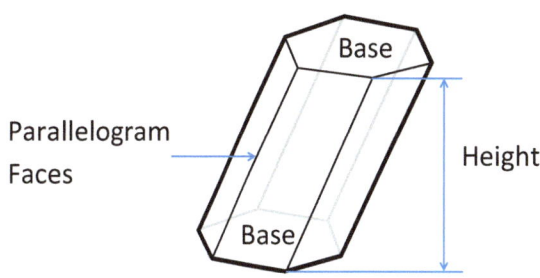

Figure 8-22: Right Pentagonal Prism

Figure 8-23: Oblique Hexagonal Prism

A regular prism is a prism with regular polygonal bases. The edge lengths of the shape are equal. If the bases are not regular polygons, then the prism is an irregular prism. The dual polyhedron

23

of a regular right prism is a bipyramid. Both a regular prism and an irregular prism can be either a right prism or an oblique prism. The figures below show a regular prism and an irregular prism.

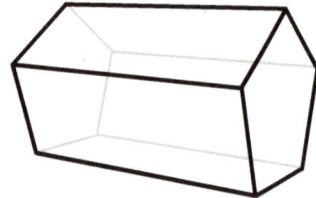

Figure 8-24: Regular Pentagonal Prism

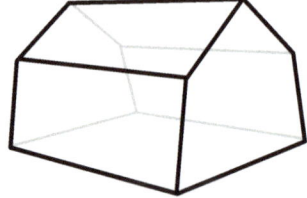

Figure 8-25: Irregular Pentagonal Prism

The surface area formula of a prism can be written as: Ph + 2B. P is the perimeter of the base, h is the height of the prism, and B is the area of the base. Ph is the lateral surface area. The volume formula of a prism can be written as: Bh. It is the product of the base area and the height or perpendicular distance between the two bases. The figures below show two sample prisms. The first figure is a cubic prism and the second figure is a triangular prism.

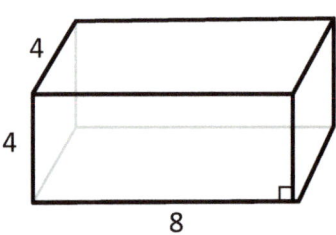

Figure 8-26: Rectangular Cubic Prism

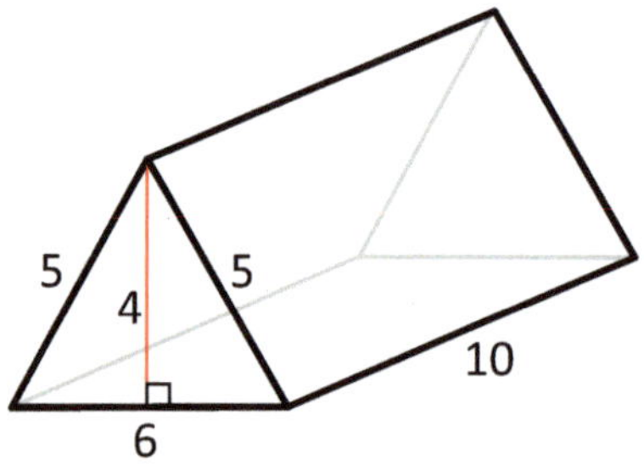

Figure 8-27: Isosceles Triangular Prism

The surface area of the cubic prism is calculated as: 24 x 4 + 2 x 32. The result of the formula is 160 square units. The volume of the cubic prism is calculated as: 32 x 4. The result of the formula is 128 cubic units. The surface area of the triangular prism is 16 x 10 + 2 x 12. The result of the formula is 184 square units. The volume of the triangular prism is calculated as: 12 x 10. The result of the formula is 120 cubic units.

A rectangular prism, with all lateral faces being rectangles, is also called a cuboid. An equilateral square prism is called a cube. A prism with all lateral faces being rectangles or squares is an orthoprism. A prism with all lateral faces being triangles is an antiprism.

8-7 Antiprisms

An antiprism is a polyhedron with two identical *n*-sided polygonal bases and 2*n* other triangular lateral faces connecting the corresponding sides of the bases. Each of the triangular lateral faces is congruent. The two bases have the same size and shape. Prisms are named according to the shape of their base, so an antiprism with a hexagon base is called a hexagonal antiprism. The difference between a prism and an antiprism is how the bases are connected. In an antiprism, one of the bases is rotated by an angle of $180°/n$. For example, if the base is a hexagon, then one of the bases is rotated by 30 degrees. The triangles are positioned so that each triangle shares one edge with a base and the opposite vertex with the other base, and its other two edges with a triangle on either side of it. The band of triangular lateral faces alternates around the bases, so that the next triangle will be rotated 180 degrees from the previous triangle. The figures below show a hexagonal prism and a hexagonal antiprism.

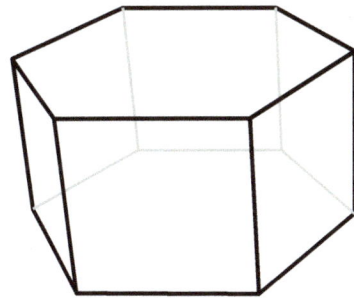

Figure 8-28: Hexagonal Prism

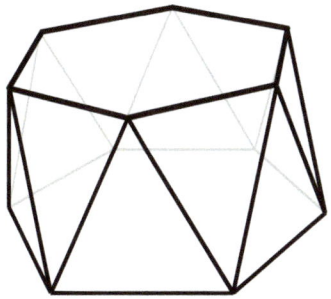

Figure 8-29: Hexagonal Antiprism

In a right antiprism, each of the triangles is an isosceles triangle. The dual polyhedrons of antiprisms are trapezohedra.

8-8 Wedges

A wedge is a polyhedron created by cutting a prism with a plane containing some edge of either base polygon, but not intersecting the other base, resulting in a half prism. A wedge has 5 faces, 9 edges, and 6 vertices. The faces consist of two triangles and three trapezoids. A wedge is a right triangular prism rotated to rest on one of its lateral rectangular faces. An oblique wedge will have its sides slanted symmetrically towards the center. The figures below show a right rectangular wedge and an oblique wedge.

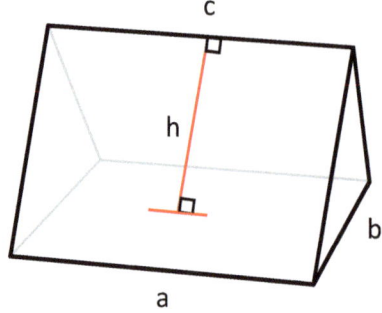

Figure 8-30: Right Rectangular Wedge

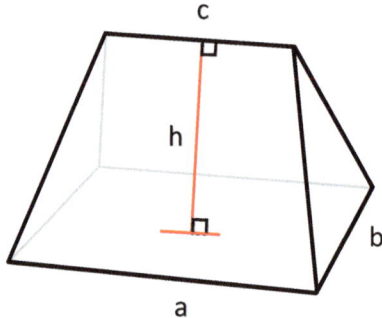

Figure 8-31: Oblique Wedge

The volume formula for a right rectangular wedge can be written as: bh (a/3 + c/6) or simplified to 1/2 abh. The base sides are labeled as a and b. The top edge is c and the height is h. Since a and c are equal, the first formula can be simplified. The volume formula for an oblique wedge can be written as: 1/6 bh (2a + c). In an oblique wedge, a and c are not equal. The figures below show wedges with sample values.

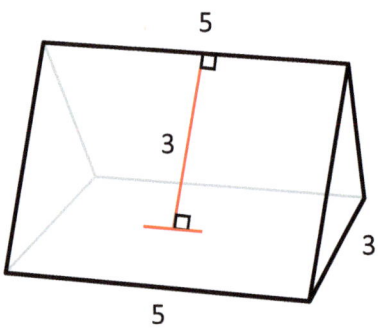

Figure 8-32: Sample Right Rectangular Wedge

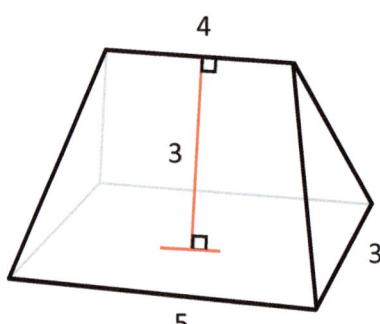

Figure 8-33: Sample Oblique Wedge

The volume of the sample right rectangular wedge is calculated as: 1/2 x 5 x 3 x 3. The result of the formula is 22.5 cubic units. The volume of the sample oblique wedge is calculated as: 1/6 x 3 x 3 (2 x 5 + 4). The result of the formula is 21 cubic units.

8-9 Toroidal Polyhedra

A toroidal polyhedron is a polyhedron with one or more holes. All of the sides are composed of polygons. A toroidal polyhedron or solid with regular polygon faces is called a Stewart toroidal polyhedron. A polyhedron is quasi-convex when it is derived by tunneling into a convex polyhedron to remove a section. It is possible to combine various polyhedra to make an infinite number of toroidal shapes. These combinations of convex polyhedra can form a ring, or multiple rings, often interconnected, and are usually highly symmetrical. The figures below show toroidal polyhedra.

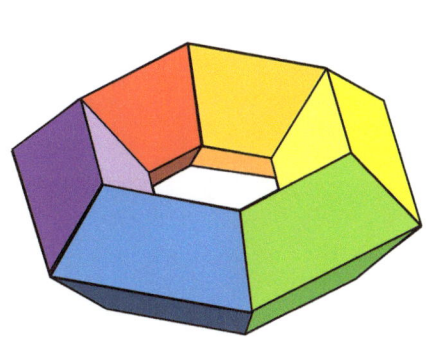

Figure 8-34: Toroidal Hexagonal Ring

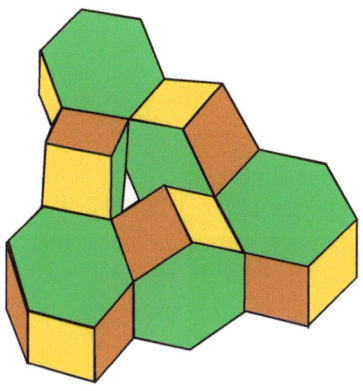

Figure 8-35: Toroidal Prismatic Triangle

Figure 8-36: Toroidal Truncated Cube

Figure 8-37: Toroidal Cubic Ring

The first figure above is created by attaching 24 congruent isosceles trapezoid faces into a ring. The second figure above is a ring of hexagonal prisms. The third figure above shows a quasi-convex truncated cube with tunneling and a cube-shaped region removed from its center. The fourth figure above shows truncated cubes combined together to make a toroidal ring.

• 8-10 Compound Polyhedra

A <mark>compound polyhedron</mark> is a polyhedron composed of a number of interpenetrating polyhedra, either of the same of several different types, which share a common center. Compound polyhedra usually have visually attractive symmetric properties. Similar to toroidal polyhedra, it is possible to combine various polyhedra to make an infinite number of compound polyhedra shapes. The figures below show compound polyhedra.

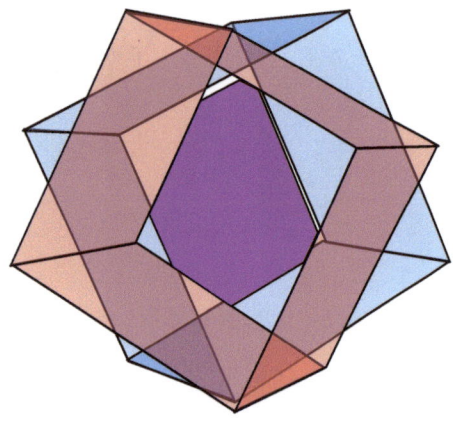

Figure 8-38: Octagrammic Prism

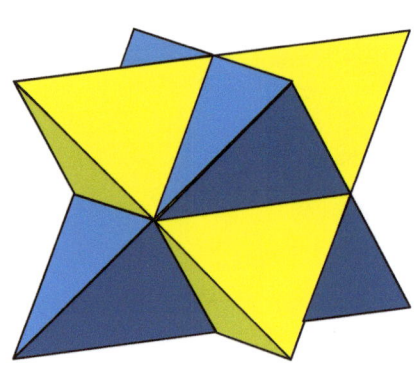

Figure 8-39: Stella Octangula

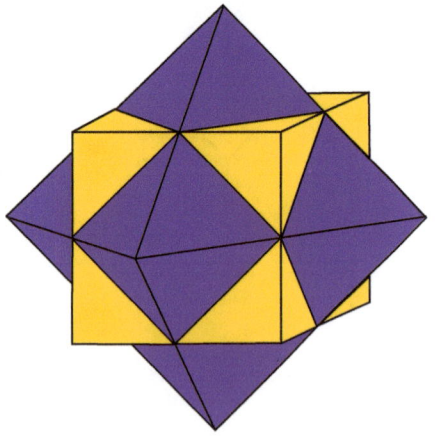

Figure 8-40: Cube and Octahedron

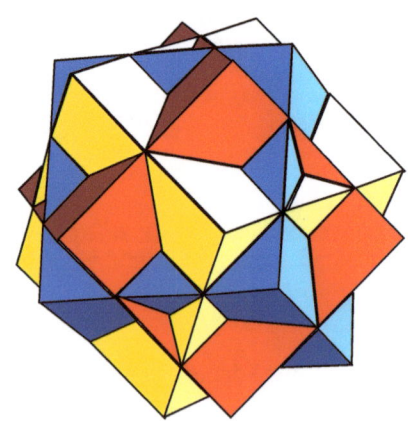

Figure 8-41: Three Cubes

The first figure above combines two cubes, one rotated 45 degrees with respect to the other, producing an octagrammic prism. The second figure above combines two interpenetrating tetrahedra, producing a stella octangula. The third figure above is a combination of a cube and an octahedron. The fourth figure above is a combination of three cubes, each rotated 45 degrees with respect to the other.

8-11 Summary

A pyramid is a polyhedron in the form of a conic solid. A conic solid is a three-dimensional geometric shape bounded by a plane base and the surface formed by line segments connecting the perimeter of the base to a common point, or apex, outside the plane of the base. A conic solid with a polygon base is a pyramid. A conic solid with a circular base is a cone. The base of a pyramid can be any polygon shape. The sides of the pyramid, called lateral faces, are always triangular shaped. Pyramids can be classified based on the location of the vertex and by the shape of the base polygon. Pyramids can be right or oblique. The shape of the base polygon of a pyramid can be regular or irregular.

A bipyramid or dipyramid is a polyhedron created by attaching two pyramids symmetrically base to base. The resulting shape has two apices. A line, such as the altitude, passing through the two apices will intersect the base plane at the center of the bases and at a perpendicular angle. The distance from each apex to the base is equal.

A trapezohedron or antidipryramid is a polyhedron with faces composed of congruent kites. A trapazohedron is similar to a bipyramid, but instead of triangular sides, it has kite sides. However, instead of attaching the two halves symmetrically like in a bipyramid, the two halves of a trapezohedron are slightly off set to allow the vertices to match up. Trapezonedra are the dual polyhedra of the Archimedean antiprisms.

A frustum is a polyhedron created by cutting off the top of a conic solid with a plane parallel to the base of the solid. The portion between the top plane and the base plane is the frustum. If the original pyramid was a right pyramid, the associated frustum is a right frustum. An axis through the center of the upper base and the lower base intersects the bottom base perpendicularly. If the original pyramid was oblique, the associated frustum is oblique. The axis does not intersect the base perpendicularly.

A prism is a polyhedron with two identical *n*-sided polygonal bases and *n* other parallelogram lateral faces connecting the corresponding sides of the bases. The two bases have the same size and shape. All cross-sections of the prism parallel to a base are the same. Prisms are named according to the shape of their base or cross-section, so a prism with a triangular base is called a triangular prism.

An antiprism is a polyhedron with two identical *n*-sided polygonal bases and 2*n* other triangular lateral faces connecting the corresponding sides of the bases. Each of the triangular lateral faces is congruent. The two bases have the same size and shape. Prisms are named according to the shape of their base, so an antiprism with a hexagon base is called a hexagonal antiprism. The difference between a prism and an antiprism is how the bases are connected.

A wedge is a polyhedron created by cutting a prism with a plane containing some edge of either base polygon, but not intersecting the other base, resulting in a half prism. A wedge has 5 faces, 9 edges, and 6 vertices. The faces consist of two triangles and three trapezoids. A wedge is a right triangular prism rotated to rest on one of its lateral rectangular faces. An oblique wedge will have its sides slanted symmetrically towards the center.

A toroidal polyhedron is a polyhedron with one or more holes. All of the sides are composed of polygons. A toroidal polyhedron or solid with regular polygon faces is called a Stewart toroidal polyhedron. A polyhedron is quasi-convex when it is derived by tunneling into a convex polyhedron to remove a section. It is possible to combine various polyhedra to make an infinite number of toroidal shapes. These combinations of convex polyhedral can form a ring, or multiple rings, often interconnected, and are usually highly symmetrical.

A compound polyhedron is a polyhedron composed of a number of interpenetrating polyhedral, either of the same of several different types, which share a common center. Compound polyhedra usually have visually attractive symmetric properties. Similar to toroidal polyhedra, it is possible to combine various polyhedra to make an infinite number of compound polyhedra shapes.

CHAPTER 8 — Chapter Test

Grading Scale: One point for each correct answer.

Excellent = 68-75, Good = 60-67, Average = 53-59, Fair = 45-52, Poor = 0-44

8-2 Pyramids

Calculate the surface area (square units) and volume (cubic units) of a square pyramid. The formulas are as follows: Surface Area = $2bs + b^2$ and Volume = $1/3\ b^2h$.

1. Base = 4, Slant Height = 2.5, Height = 1.5   _____   _____
2. Base = 6, Slant Height = 3.75, Height = 2.25   _____   _____
3. Base = 8, Slant Height = 5, Height = 3   _____   _____
4. Base = 10, Slant Height = 6.25, Height = 3.75 _____   _____
5. Base = 12, Slant Height = 7.5, Height = 4.5   _____   _____
6. Base = 16, Slant Height = 10, Height = 6   _____   _____

8-3 Bipyramids

Calculate the surface area (square units) and volume (cubic units) of a square bipyramid. The formulas are as follows: Surface Area = $4bs$ and Volume = $2/3\ b^2h$. The height of a bipyramid is the distance from the base to an apex.

1. Base = 4, Slant Height = 2.5, Height = 1.5   _____   _____
2. Base = 6, Slant Height = 3.75, Height = 2.25   _____   _____
3. Base = 8, Slant Height = 5, Height = 3   _____   _____
4. Base = 10, Slant Height = 6.25, Height = 3.75 _____   _____
5. Base = 12, Slant Height = 7.5, Height = 4.5   _____   _____
6. Base = 16, Slant Height = 10, Height = 6   _____   _____

8-4 Trapezohedra

Mark as True or False.

1. A trapezahedron is also called an antidipryramid.   ____
2. A trapazohedron has trapezoid sides.   ____
3. The *n*-gon part of the name refer to the number of faces.   ____
4. The center of mass of a trapezohedron is at the center of the base.   ____

8-5 Frusta

Calculate the surface area (square units) and volume (cubic units) of a square pyramidal frusta. The formulas are as follows: Surface Area = 1/2(p1+p2)s and Volume = 1/3 h(a1+a2 + √(a1a2)). P1 = Perimeter of base 1. P2 = Perimeter of base 2. A1 = Area of base 1. A2 = Area of base 2.

1. Base 1 side = 4, Base 2 side = 1, Slant Height = 3, Height = 2.5   _____ _____
2. Base 1 side = 8, Base 2 side = 2, Slant Height = 6, Height = 5   _____ _____
3. Base 1 side = 12, Base 2 side = 3, Slant Height = 9, Height = 7.5   _____ _____
4. Base 1 side = 16, Base 2 side = 4, Slant Height = 12, Height = 10   _____ _____
5. Base 1 side = 20, Base 2 side = 5, Slant Height = 15, Height = 12.5   _____ _____
6. Base 1 side = 24, Base 2 side = 6, Slant Height = 18, Height = 15   _____ _____

8-6 Prisms

Calculate the surface area (square units) and volume (cubic units) of a cubic prism. The formulas are as follows: Surface Area = Ph + 2b and Volume = Bh. P = perimeter of base. H = height. B = area of base.

1. Length = 8, Width = 4, Height = 4   _____ _____
2. Length = 12, Width = 6, Height = 6   _____ _____
3. Length = 16, Width = 8, Height = 8   _____ _____
4. Length = 20, Width = 10, Height = 10   _____ _____

Calculate the surface area (square units) and volume (cubic units) of an isosceles triangular prism. The formulas are as follows: Surface Area = Ph + 2b and Volume = Bh. P = perimeter of base. H = height. B = area of base.

5. Triangle sides 5, 5, 6; Height = 4; Length = 10   _____ _____
6. Triangle sides 7.5, 7.5, 9; Height = 6; Length = 15   _____ _____
7. Triangle sides 10, 10, 12; Height = 8; Length = 20   _____ _____
8. Triangle sides 12.5, 12.5, 15; Height = 10; Length = 25   _____ _____

8-7 Antiprisms

Mark as True or False.

1. An antiprism has two identical polygonal bases and triangular lateral faces.   ____
2. Each of the triangular lateral faces is unique.   ____
3. In a pentagonal antiprism, then one of the bases is rotated by 45 degrees.   ____
4. The band of faces alternates around the bases, 90 degrees from the previous face.   ____

8-8 Wedges

Calculate the volume of a right rectangular wedge. The formula is as follows: 1/2 abh.
A = base a. B = base b. C = top edge. H = height.

1. Base a = 5, Base b = 3, Top edge = 5, Height = 3 _____
2. Base a = 7.5, Base b = 4.5, Top edge = 7.5, Height = 4.5 _____
3. Base a = 10, Base b = 6, Top edge = 10, Height = 6 _____
4. Base a = 12.5, Base b = 7.5, Top edge = 12.5, Height = 7.5 _____

Calculate the volume of an oblique wedge. The formula is as follows: 1/2 abh. A = base a.
B = base b. C = top edge. H = height.

5. Base a = 5, Base b = 3, Top edge = 4, Height = 3 _____
6. Base a = 7.5, Base b = 4.5, Top edge = 6, Height = 4.5 _____
7. Base a = 10, Base b = 6, Top edge = 8, Height = 6 _____
8. Base a = 12.5, Base b = 7.5, Top edge = 10, Height = 7.5 _____

8-9 Toroidal Polyhedra

Mark as True or False.

1. A toroidal polyhedron is a polyhedron with at least two. _____
2. All of the sides are composed of regular polygons. _____
3. There are an infinite number of toroidal shapes. _____
4. These combinations of convex polyhedra can form a ring, or multiple rings. _____

8-10 Compound Polyhedra

Mark as True or False.

1. A compound polyhedron is composed of interpenetrating polyhedra. _____
2. The polyhedra do not share a common center. _____
3. There are an infinite number of compound polyhedra shapes. _____

CHAPTER 9 Two Dimensional Non-polytopes

9-1 Introduction

Two dimensional non-polytopes are geometric objects with two dimensions that do not have straight sides. The sides of non-polytopes are curved. This section will describe the following shapes: circle, ellipse, parabola, hyperbola, circular sector, circular segment, circular ring, circular ring sector, stadium, and oval. The first four objects can be generated based on the construction of conic sections. The last six objects are derived from a circle or ellipse.

9-2 Conic Sections

A conic section is a curve resulting from the intersection of a right circular cone and a plane. A cone, when describing conic sections, means a double cone, which are two cones placed apex to apex. Each of the two cones is called a nappe. A doubly infinite cone, or double cone, is the union of any set of straight lines that pass through a common apex point, and therefore extends symmetrically on both sides of the apex. This kind of cone does not have a bounding base and extends to infinity. The boundary of a double cone is a conical surface, and the intersection of a plane with this surface is a conic section.

The conical surface of a double cone is generated by rotating a line in the Y-Z plane about the Z axis with the vertex at the origin. The Z axis is the axis of the cones. A cone is determined by its vertex angle. The figures below show the construction of a double cone and the resulting shape.

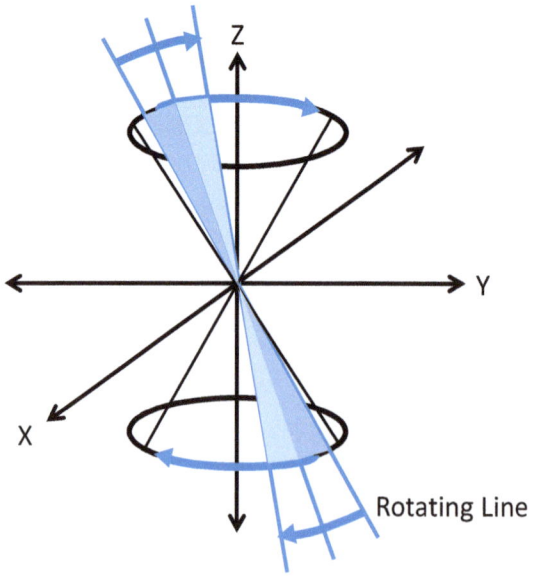

Figure 9-1: Double Cone Construction

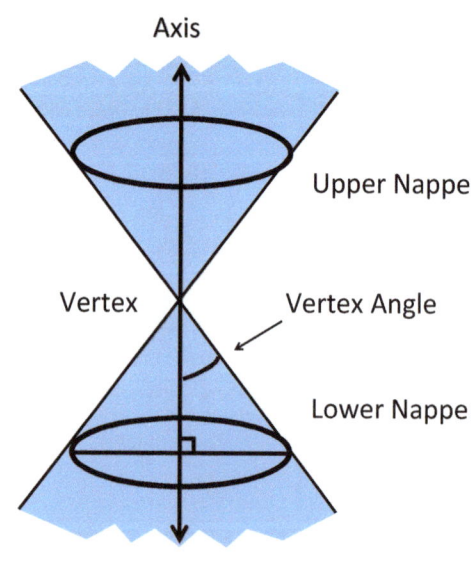

Figure 9-2: Double Cone

A cone is a surface of revolution generated by an oblique line that rotates around a fixed point, at a fixed angle from the axis, with both the line and axis passing through that fixed point. This oblique line is referred to as the generator. The generated cone is hollow, not a solid object. If the base is a circle, the shape is a circular cone. If the axis from the center of the base to the vertex is perpendicular to the base, then it is a right cone; otherwise it is an oblique cone. The curved lateral surface of the cone is the nappe.

9-3 Types of Conics

By changing the angle of intersection of the plane and the cone, four basic types of conics can be produced. These four types of conics are circles, ellipses, parabolas, and hyperbolas. When the plane intersects the vertex of the cone, the resulting conic is called a degenerate conic. Degenerate conics include a point, a line, and two intersecting lines. The circle and ellipse are closed curves, while the parabola and hyperbola are unbounded curves. The figures below show the four conic sections and the four degenerate conics.

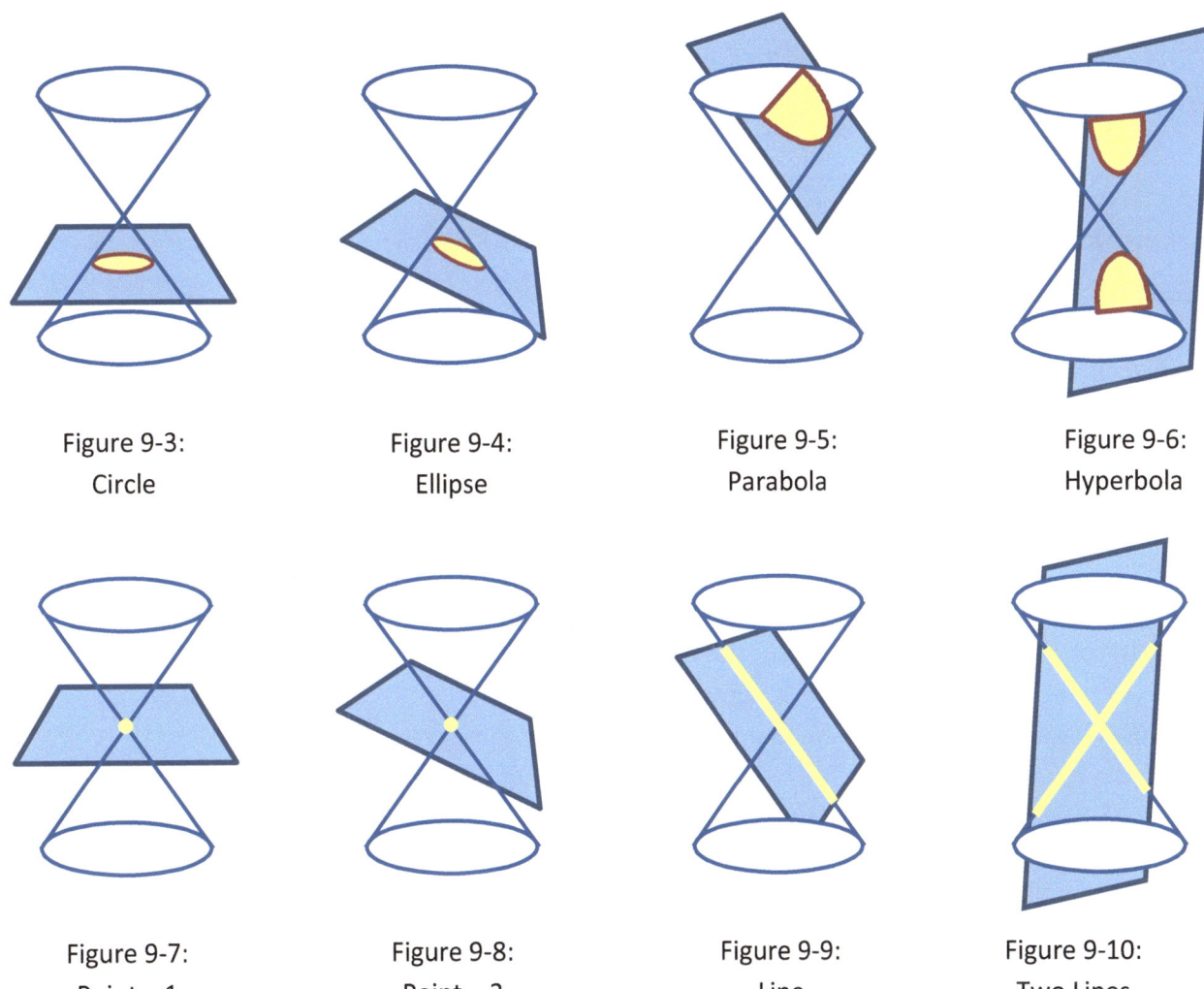

Figure 9-3: Circle

Figure 9-4: Ellipse

Figure 9-5: Parabola

Figure 9-6: Hyperbola

Figure 9-7: Point – 1

Figure 9-8: Point – 2

Figure 9-9: Line

Figure 9-10: Two Lines

A plane intersecting a double cone at a right angle to the axis creates a cross section in the shape of a circle. As the plane moves closer to the vertex, the size of the circle becomes small. As the plane moves farther from the vertex, the circle becomes larger. If the plane intersects the vertex, the circle degenerates to become a point.

A plane intersecting one nappe, but not at a right angle or parallel to the side of the cone, creates a cross section in the shape of an ellipse. As the plane move closer to the vertex, the size of the ellipse becomes smaller. As the plane move farther from the vertex, the size of the ellipse becomes larger. If the plane intersects the vertex, the ellipse degenerates to become a point.

A plane intersecting one nappe parallel to the side of the cone creates a cross section in the shape of a parabola. As the plane moves closer to the vertex, the size of the parabola becomes narrower. As the plane move farther from the vertex, the size of the parabola becomes wider. If the plane intersects the vertex, the parabola degenerates to become a line.

A plane intersecting both nappes creates a hyperbola. As the plane move closer to the vertex, the branches of the hyperbola become narrower and closer together. As the plane move farther from the vertex, the branches of the hyperbola become wider and farther apart. If the plane intersects the vertex, the hyperbola degenerates to become two intersecting lines.

9-4 Circles

A circle is the set of all points in a plane at an equal distance from a fixed point. This set of points forms a continuous closed curved line. The fixed point is the center of the circle. The distance from the center of the circle to a point on the circle is called the radius. A chord is a line segment whose end points lie on the circle. A diameter is a chord that passes through the center of the circle. A radius equals one half of a diameter. A tangent line is a line that intersects a circle in exactly one point. The point of contact is called the point of tangency. A secant line is a line that intersects a circle in two different points. Every secant line includes a chord of a circle.

The interior of a circle is the set of all points whose distance from the center is less than the length of the radius of the circle. The exterior of a circle is the set of all points whose distance from the center is greater than the length of the radius of the circle. The area enclosed by a circle is called a disk. The area of a disk is commonly referred to as the area of a circle.

A central angle is an angle whose vertex is at the center of the circle. An arc is a curved portion of a circle. A minor arc is an arc that lies in the interior of the central angle that intercepts the arc. The measure of a minor arc is the same as the measure of its central angle. A major arc is an arc that lies in the exterior of the central angle that intercepts the arc. The degree measurement of a major arc is equal to 360 minus the measure of the central angle that determines the end points of the arc. The measure of an arc, as with angle measurement, is denoted by preceding the name of the arc by the letter "m": mAB = 45 is read as the measure of arc AB is 45. When this notation is used, the degree symbol is omitted. Congruent arcs are arcs in the same circle or congruent circles that have the same degree measure. The figures below show the lines and parts of a circle.

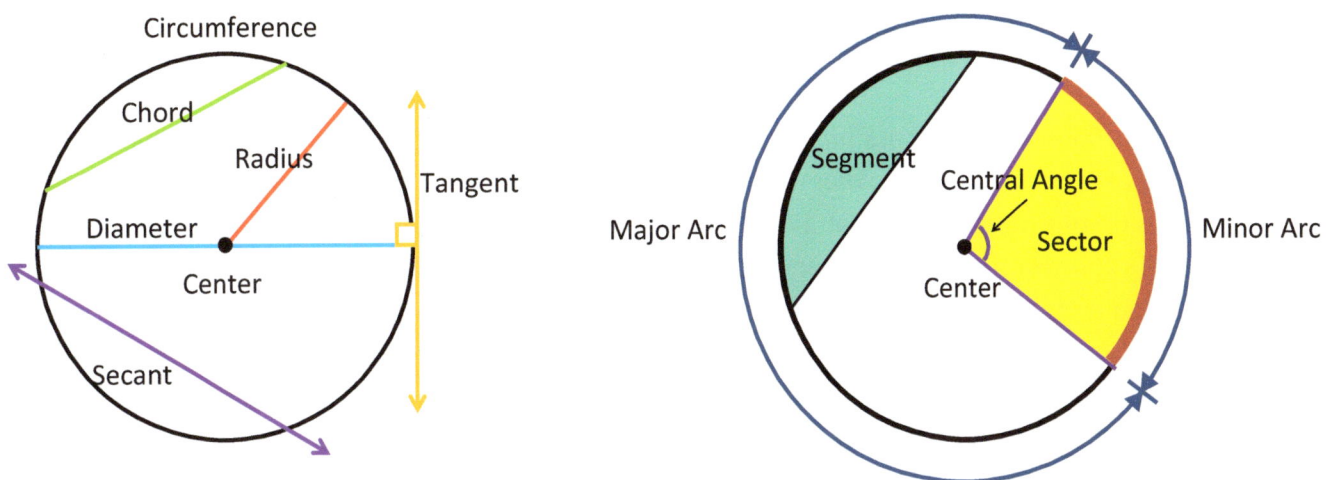

Figure 9-11: Lines of a Circle　　　　　　　　　　Figure 9-12: Parts of a Circle

The perimeter is the measurement around the outside of an object. The perimeter of a circle is called the circumference. Unlike polygons, which are made of many straight sides, a circle only has one curved side. A semicircle is half of a circle and is created by drawing a diameter. A circle has a unique relationship between the circumference and the diameter. The ratio of the circumference and diameter is constant for all circles. The value of the circumference divided by the diameter is represented by the Greek letter π, which is also spelled pi. This number is a nonrepeating, nonterminating decimal approximately equal to 3.141592654 to nine decimal places.

The formula for the circumference of a circle is: $c = \pi d$ or $2\pi r$. The formula for the area of a circle is: $A = \pi r^2$ or $(\pi/4) \times d^2$. The formula for the length of an arc based on a central angle measurement is: $(\pi/180) \times$ angle $\times$ radius. The following examples show these formulas. If the radius of a circle is 4 units, then the circumference of the circle is 25.1327 units. The area of the circle is 50.2654 units. If the central angle is 60 degrees, then the length of the arc is 4.1887 units.

The table below shows the formulas for the radius, diameter, circumference, and area of a circle.

Measurement	Formula 1	Formula 2	Formula 3
Radius (r)	$r = d/2$	$r = c/2\pi$	
Diameter (d)	$d = 2r$	$d = c/\pi$	$d = \sqrt{4A/\pi}$
Circumference (c)	$c = 2\pi r$	$c = \pi d$	$c = \sqrt{4\pi A}$
Area (A)	$A = \pi r^2$	$A = (\pi d^2)/4$	$A = c^2/4\pi$

Table 9-1: Formulas of a Circle

A circle has the maximum possible area for a given perimeter, and the minimum possible perimeter for a given area. The diameter is the longest chord of a circle. If two chords intersect, one chord divides into lengths *a* and *b* and the other chord divides into lengths *c* and *d*. This divides the chords proportionally such that ab = cd. If two chords intersect at perpendicular angles, then $a^2 + b^2 + c^2 + d^2$ equals the square of the diameter. A line perpendicular to a radius through the end point of the radius is tangent to a circle. If there are two circles, there are lines that can be tangent to both of the circles at the same time. The following figures show the four possible common tangents for two circles that do not intersect.

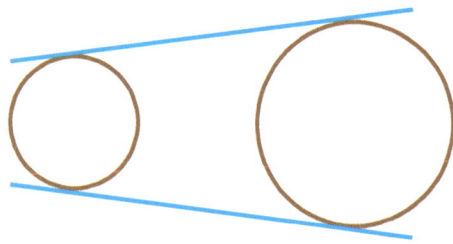

Figure 9-13: Two External Tangents

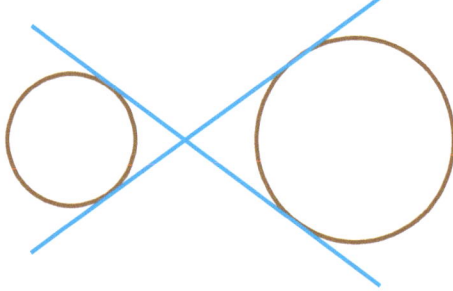

Figure 9-14: Two Internal Tangents

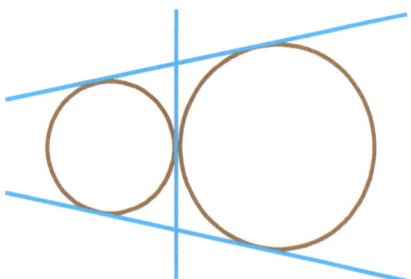

Figure 9-15: Two External, One Internal Tangents

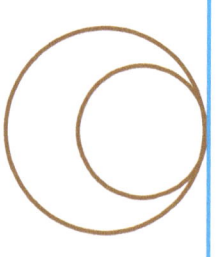

Figure 9-16: One External Tangent

A diameter line divides a circle into two equal semicircles. The perimeter of a semicircle is not half of the perimeter of a circle. The formula for the perimeter of a semicircle is: $P = \pi r + 2r$ or $r(\pi + 2)$. The formula for the area of a semicircle is: $A = (\pi r^2) / 2$. The following examples show these formulas. If the radius of a semicircle is 4 units, then the perimeter of the semicircle is 20.5663 units. The area of the semicircle is 25.1327 units.

9-5 Circular Sectors

A circular sector is part of a circle bounded by two radii and their intercepted arc. It is a wedge shaped piece of a circle. The arc is based on the central angle formed by the two radii. The formula for the perimeter of a sector is: $P = ((\pi/180) \times angle \times radius) + 2r$ or $r(2 + \pi(angle/180))$. The formula for the area of a sector is: $A = (angle \times \pi r^2) / 360$. The angle is measured in degrees. The figure below shows a circular sector.

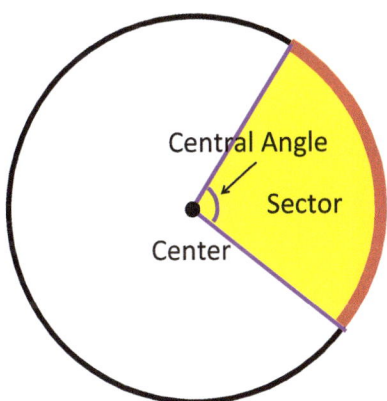

Figure 9-17: Circular Sector

9-6 Circular Segments

A circular segment is part of a circle bounded by a chord and its associated arc. The formula for the perimeter of a segment is: $P = ((\pi/180) \times angle \times radius) + c$. The area of the segment is equal to the area of the sector minus the area of the triangular section. The formula for the area of a segment is: A = ((angle x πr²) / 360) – (cd/2). The first figure below shows a circular segment. The second figure below shows a detailed drawing of a circular segment with labeled parts.

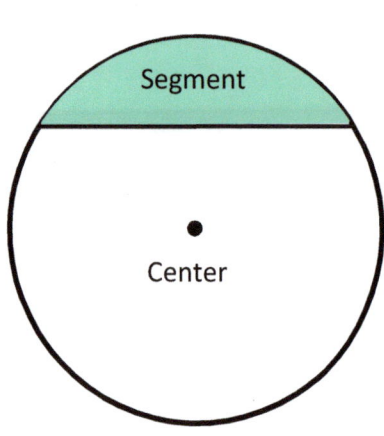

Figure 9-18: Circular Segment

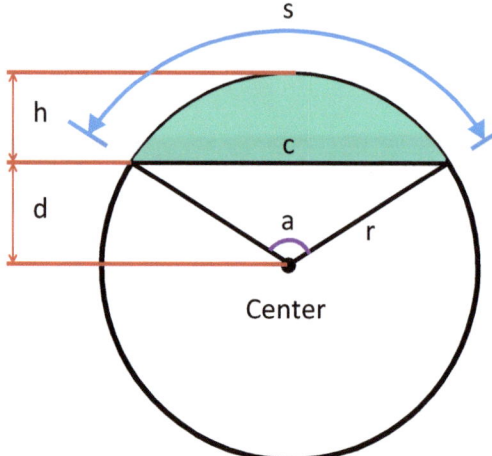

Figure 9-19: Circular Segment - Detail

There are several other formulas for finding the values of the parts of a circular segment. The formula for the length of the arc is: s = ((π/180)(ar)) or ((a/360)(2πr). The formula for the height of the segment is: h = r-√(r²- (c²/4)) or √((r²- (r/2)²). The formula for the height of the triangular section is: d = 1/2√(4r²- c²). The formula for the chord length is: c = 2√((h(2r – h)) or 2√(r²-d²). The formula for the radius is: r = h + d or (c² + 4h²) / 8h or (h/2) + (c²/8h). The formula for the central angle is: a = 360s/2πr.

• 9-7 Circular Rings

A circular ring or annulus is a ring shaped object and is the region lying between two concentric coplanar circles. There are two methods for calculating the area of a circular ring. The first method is calculated by subtracting the area of the smaller circle from the area of the larger circle. The formula for the area of a circular ring is: A = π (r1² – r2²) or πr1² – πr2² or (π/4)(r1² – r2²). The second method is calculated by multiplying the average radius of the two circles (r3) by the width of the ring (w). The formula for the area of a circular ring is: A = 2πr3w. The figures below show circular rings.

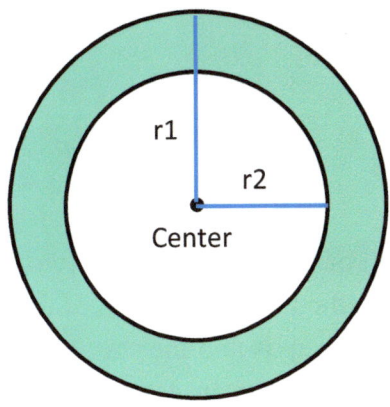
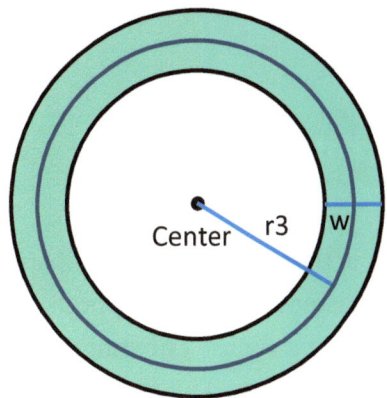

Figure 9-20: Circular Ring – Two Radii   Figure 9:21: Circular Ring – Average Radius

9-8 Circular Ring Sectors

A circular ring sector or annulus sector is part of a ring shaped object and is part of the region lying between two concentric coplanar circles. There are two methods for calculating the area of a circular ring sector. The first method is calculated by subtracting the area of the smaller circular sector from the area of the larger circular sector. The formula for the area of a circular ring is: A = ((angle x πr1^2) / 360) – ((angle x πr2^2) / 360). The second method is calculated by multiplying angle by the average radius of the two circles (r3) by the width of the ring (w). The formula for the area of a circular ring sector is: A = a(180/π)r3w. The figures below show circular ring sectors.

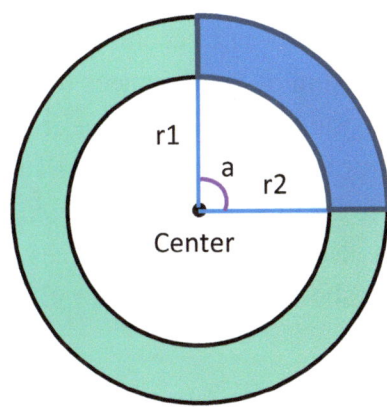
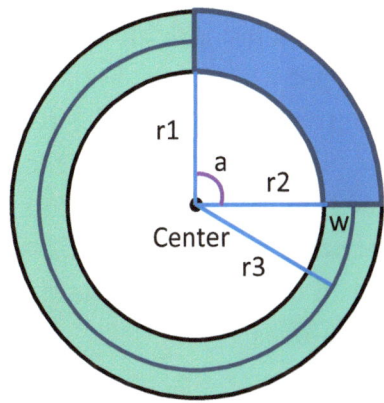

Figure 9-22: Circular Ring Sector – Two Radii   Figure 9-23: Circular Ring Sector – Average Radius

9-9 Ellipses

An ellipse is the set of all points in a plane in which the sum of the distances from two fixed points is constant. This set of points forms a continuous closed curved line. Each of the two points is called a focus. The longest diameter is called the major axis and the shortest diameter is called the minor axis. The major axis and minor axis are perpendicular bisectors of each other. The figures below show a general ellipse and an ellipse with points.

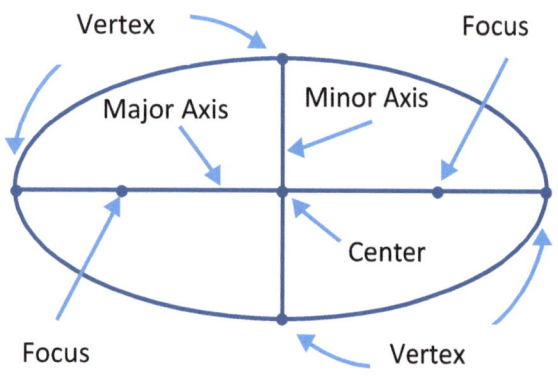

Figure 9-24: Ellipse

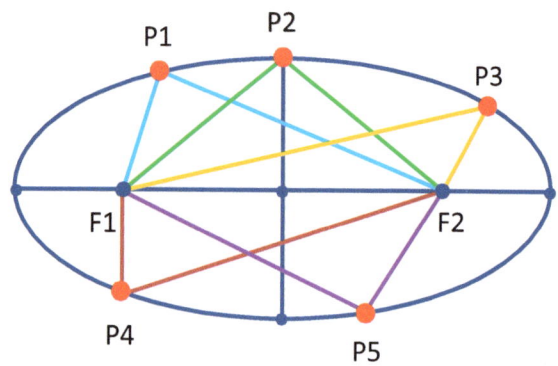

Figure 9-25: Ellipse with Points

The first figure above shows an ellipse. Each axis is divided into two equal sections separated by the center. The major axis is made up of two symmetrical semi-major axes and the minor axis is made up of two symmetrical semi-minor axes. At the end of each axis, where it meets the ellipse, is a vertex. An ellipse is defined by two points, each called a focus. The second figure above shows an ellipse with points. For any point on the ellipse, the sum of the distances to the focus points is constant. The sum of the distance from a focus to a point plus the distance from that point to the other focus is the same for every point. For example, the distance from F1 to P1 plus the distance from P1 to F2 is equal to the distance from F1 to P2 plus the distance from P2 to F2. This can be written as F1P1 + P1F2 = F1P2 + P2F2. This is also equal to the distances for P3, P4, P5, and all other points on the ellipse. The figures below show an ellipse with ratios and an ellipse with a focus.

42

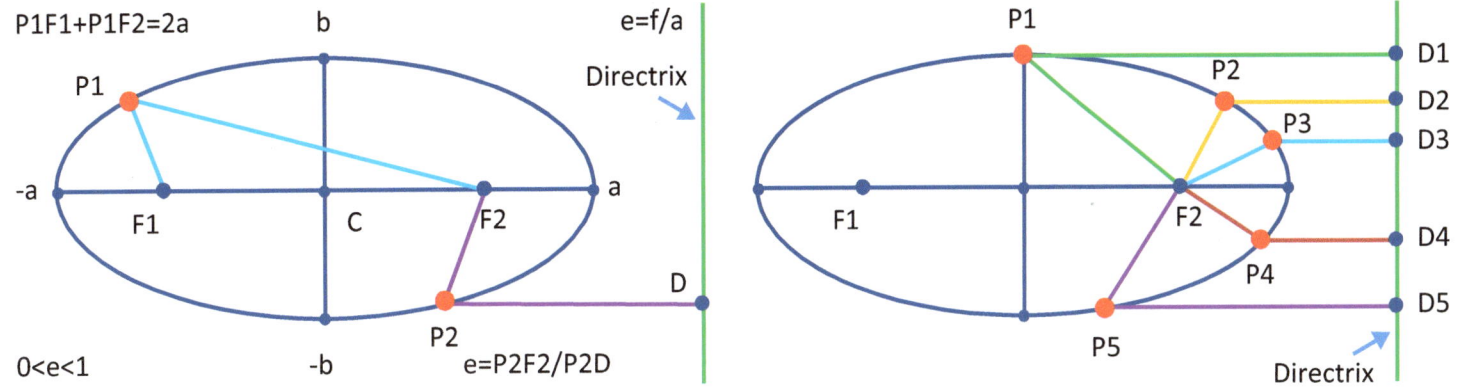

Figure 9-26: Ellipse Ratios

Figure 9-27: Ellipse Focus

The first figure above shows the ellipse ratios. The ratios of an ellipse define its shape and measurements. The sum of the distance from a focus to any point plus the distance from that point to the other focus is equal to the length of the major axis, or twice the distance from the center to a vertex of the major axis. For example, P1F1 + P1F2 = 2a. A <mark>directrix</mark> is a line outside of an ellipse that is parallel to the minor axis. The distance from the major axis vertex to the directrix must be greater than the distance from the major axis vertex to the focus. The <mark>eccentricity</mark> is the ratio of minor axis length to major axis length. The eccentricity is equal to the ratio of the distance from a focus to a point divided by the horizontal distance to the directix. For example, e = P2F2 / P2D. The eccentricity is also equal to the ratio of the distance from the center to a focus divided by the distance from the center to a vertex of the major axis. For example, e = CF2 / Ca. The eccentricity will have a value greater than zero and less than one.

The length of the major axis and minor axis can be calculated. Using the first figure above, the length of the major axis is P1F1 + P1F2. The length of the minor axis is $\sqrt{(P1F1 + P1F2)^2 + F1F2)}$. The location of the foci can be calculated. The distance of each focus from the center on the major axis is calculated as: $F = \sqrt{a^2 - b^2}$.

The second figure above shows an ellipse focal point. The location of a focus is defined as a ratio of the distance of the focus from a point divided by the distance of that point from a directrix. This constant ratio defines a focus point. For example, the distance from F2 to P1 divided by the distance from P1 to D1 is an equal ratio to the distance from F2 to P2 divided by the distance from P2 to D2. This can be written as F2P1 / P1D1 = F2P2 / P2D2. This is also equal to the ratios for P3, P4, P5 and all other points on the ellipse.

A circle is a special case of an ellipse in which the major axis and minor axis have the equal lengths. As a result, the area and circumference formulas for a circle and ellipse are similar. The area of an ellipse can be written as: $A = \pi ab$, where a and b are one half of the major and minor axis lengths. The area of a circle is: $A = \pi r^2$. If the radius of a circle is labeled as a, then the area of a circle can be written as: $A = \pi a^2$ or πaa. The circumference of an ellipse can be written as: $C = \pi(a + b)$ or $C = 2\pi\sqrt{((a^2 + b2)/2)}$. The results of the formulas are approximate values because of eccentricity of ellipses. The circumference of a circle is $2\pi r$. If the radius of a circle is labeled as a, then the circumference can be written as $2\pi a$ or $\pi(a + a)$.

- ## 9-10 Parabolas

A parabola is the set of all points in a plane in that are equidistance from a fixed line and a fixed point not on the line. The line is the directrix and the point is the focus. The vertex is a point on the parabola halfway between the directrix and the focus. The axis is a line passing through the focus and vertex. A parabola is symmetrical along its axis. The figures below show a general parabola and a parabola with a focus.

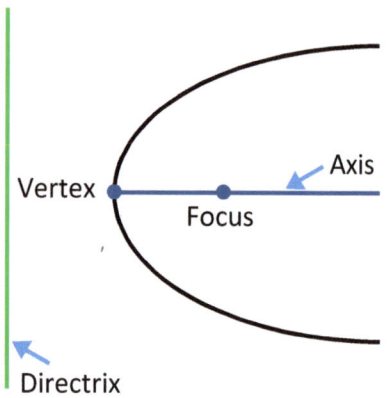

Figure 9-28: Parabola

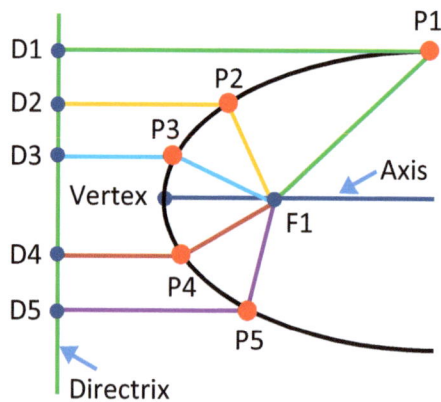

Figure 9-29: Parabola Focus

The distance from the focus to a point is equal to the distance from that point to the directrix. For example, the distance from F1 to P1 is equal to the distance from P1 to D1, or F1P1 = P1D1. This is also true for P2, P3, P4, P5, and all other points on the parabola. The area of a parabola can be calculated using two formulas. The area can be calculated as 2/3 of the base times the height and written as: A = 2/3(bh). The area can also be calculated as 2/3 of the circumscribed parallelogram formed by the chord of the parabola and a tangent of the parabola. The figures below show the area of a parabola using the base times height and parallelogram formulas.

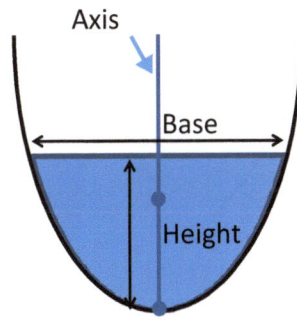

Figure 9-30: Parabola – Formula 1

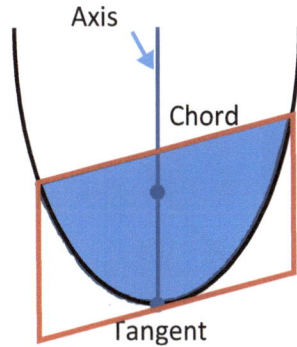

Figure 9-31: Parabola – Formula 2

• 9-11 Hyperbolas

A hyperbola is the set of all points in a plane in that the difference between each of two fixed points and any point on the hyperbola is constant. The two fixed points are the foci of the hyperbola. The transverse axis is a line passing through the foci. A hyperbola is symmetrical along its axis. A hyperbola consisted of two disconnected curves or branches. The midpoint between the branches is the center. The points on the two branches closest to the center are the vertices. The conjugate axis is perpendicular to the transverse axis and passes through the center. Asymptote lines form the boundary of the hyperbola curve. The hyperbola approaches the asymptote as the distance from the center increases, but it does not intersect it. The figure below shows a general hyperbola.

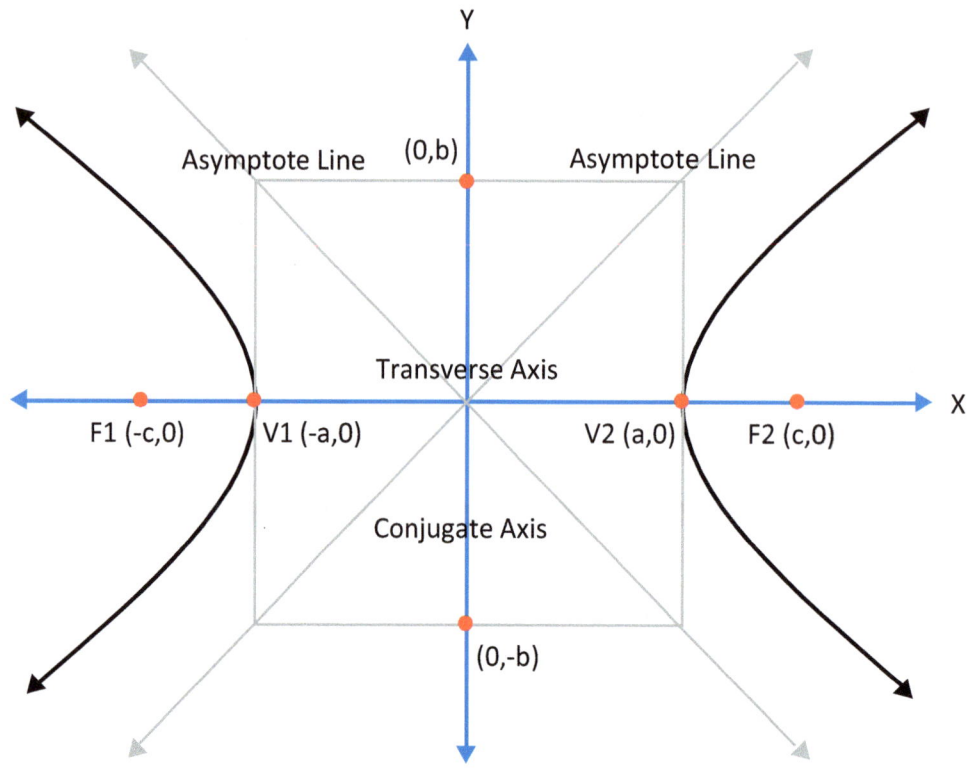

Figure 9-32: Hyperbola

The distance from a focus to a point on the hyperbola minus the distance from that point to the other focus is equal for all points on the hyperbola. For example, the distance from F1 to P1 minus the distance from P1 to F2 is equal to the distance from F1 to P2 minus the distance from P2 to F2, or F1P1 − P1F2 = F1P2 − P2F2. This is also true for P3, P4, P5, and all other points on the hyperbola. The figure below shows a hyperbola with foci.

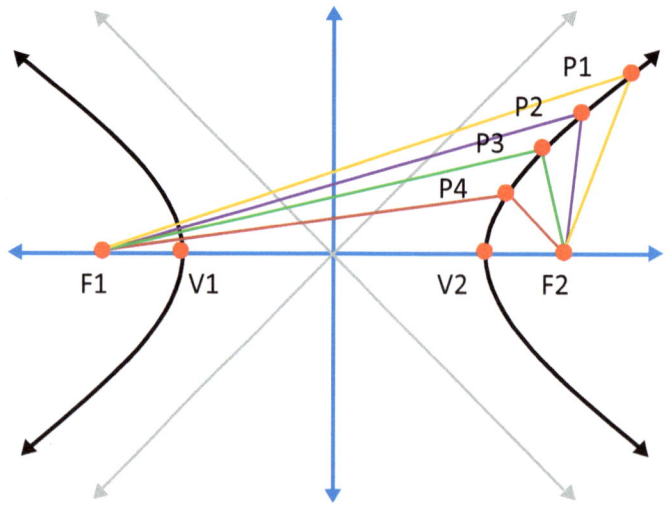

Figure 9-33: Hyperbola Foci

- ## 9-12 Stadiums

A stadium is a shape consisting of a rectangle with semicircles attached to two opposite ends. The figures below show a stadium and a stadium with sample values.

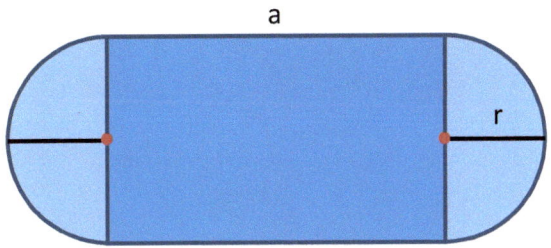

Figure 9-34: Stadium   Figure 9-35: Stadium Values

The area of a stadium is written as: $A = \pi r^2 + 2ra$. Using the sample values, the area is 22.0685 square units. The perimeter of a stadium is written as: $P = 2(\pi r + a)$. The perimeter is 19.4247 units. The side length of a stadium is written as: $a = (P/2) - (\pi r)$ or $a = (A - (\pi r^2)) / (2r)$. The side length is 5 units. The radius of a stadium is written as: $r = (P/2\pi) - (a/\pi)$. The radius is 1.5 units.

- ## 9-13 Ovals

An oval is a shape consisting of a closed curve that may or may not be symmetrical. It is unlike a circle or ellipse and does not have a precise mathematical definition. Oval means egg shaped. One type of oval is constructed by overlapping two unequal circles and then adding tangential curves to smooth out the shape. The radius of the larger circle (R) is larger than the radius of the smaller circle (r). The distance between the two radii (a) is greater than the difference in the radii measurements. The radius of the joining circle (p), which creates the tangential curves, is calculated as: $p = (a^2 + R^2 - r^2) / ((2(R - r))$. The center of the joining circle (0,y) is calculated as: $y = ((R - r)^2 - a^2) / ((2(R - r))$. The area of an oval uses the same formula as a stadium and is written as: $A = \pi r^2 + 2ra$. Other types of ovals can be constructed with tangent circles and more complex designs. The figures below show an oval and an oval construction.

Figure 9-36: Oval

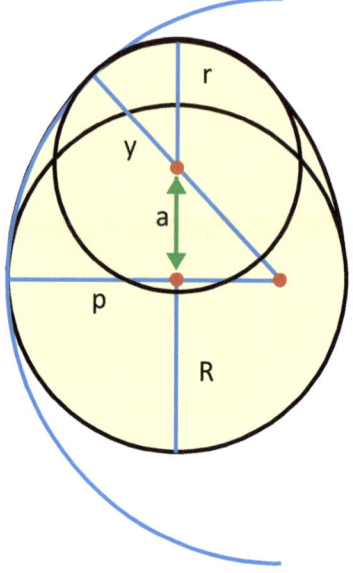

Figure 9-37: Oval Construction

9-14 Summary

Two dimensional non-polytopes are geometric objects with two dimensions that do not have straight sides. The sides of non-polytopes are curved. A conic section is a curve resulting from the intersection of a right circular cone and a plane. A cone, when describing conic sections, means a double cone, which are two cones placed apex to apex. Each of the two cones is called a nappe. A doubly infinite cone, or double cone, is the union of any set of straight lines that pass through a common apex point, and therefore extends symmetrically on both sides of the apex. This kind of cone does not have a bounding base and extends to infinity. The boundary of a double cone is a conical surface, and the intersection of a plane with this surface is a conic section.

By changing the angle of intersection of the plane and the cone, four basic types of conics can be produced. These four types of conics are circles, ellipses, parabolas, and hyperbolas. When the plane intersects the vertex of the cone, the resulting conic is called a degenerate conic. Degenerate conics include a point, a line, and two intersecting lines. The circle and ellipse are closed curves, while the parabola and hyperbola are unbounded curves.

A circle is the set of all points in a plane at an equal distance from a fixed point. This set of points forms a continuous closed curved line. The fixed point is the center of the circle. The distance from the center of the circle to a point on the circle is called the radius. A chord is a line segment whose end points lie on the circle. A diameter is a chord that passes through the center of the circle. A radius equals one half of a diameter. A tangent line is a line that intersects a circle in exactly one point. The point of contact is called the point of tangency. A secant line is a line that intersects a circle in two different points. Every secant line includes a cord of a circle.

A circular sector is part of a circle bounded by two radii and their intercepted arc. It is a wedge shaped piece of a circle. The arc is based on the central angle formed by the two radii. A circular segment is part of a circle bounded by a chord and its associated arc. A circular ring or annulus is a ring shaped object and is the region lying between two concentric coplanar circles. A circular ring sector or annulus sector is part of a ring shaped object and is part of the region lying between two concentric coplanar circles.

An ellipse is the set of all points in a plane in which the sum of the distances from two fixed points is constant. This set of points forms a continuous closed curved line. Each of the two points is called a focus. The longest diameter is called the major axis and the shortest diameter is called the minor axis. The major axis and minor axis are perpendicular bisectors of each other.

A parabola is the set of all points in a plane in that are equidistance from a fixed line and a fixed point not on the line. The line is the directrix and the point is the focus. The vertex is a point on the parabola halfway between the directrix and the focus. The axis is a line passing through the focus and vertex. A parabola is symmetrical along its axis.

A hyperbola is the set of all points in a plane in that the difference between each of two fixed points and any point on the hyperbola is constant. The two fixed points are the foci of the hyperbola. The transverse axis is a line passing through the foci. A hyperbola is symmetrical along its axis. A hyperbola consisted of two disconnected curves or branches. The midpoint between the branches is the center. The points on the two branches closest to the center are the vertices. The conjugate axis is perpendicular to the transverse axis and passes through the center. Asymptote lines form the boundary of the hyperbola curve. The hyperbola approaches the asymptote as the distance from the center increases, but it does not intersect it.

A stadium is a shape consisting of a rectangle with semicircles attached to two opposite ends. An oval is a shape consisting of a closed curve that may or may not be symmetrical. It is unlike a circle or ellipse and does not have a precise mathematical definition.

CHAPTER 9

Chapter Test

Grading Scale: One point for each correct answer.

Excellent = 91-101, Good = 81-90, Average = 71-80, Fair = 61-70, Poor = 0-60

9-2 Conic Sections

Match definitions and terms.

    A = Conic Section   B = Double Cone   C = Conical Surface

    D = Cone   E = Generator   E = Nappe

1. Oblique line that rotates around a fixed point, at a fixed angle from the axis. _____
2. Union of any set of straight lines that pass through a common apex point. _____
3. Surface of revolution generated by an oblique line rotating around a fixed point. _____
4. Curve resulting from the intersection of a right circular cone and a plane. _____
5. Curved lateral surface of the cone. _____
6. Generated by rotating a line in the Y-Z plane about the Z axis. _____

9-3 Types of Conics

Match definitions and terms.

    A = Circle   B = Ellipse   C = Parabola   D = Hyperbola   E = Degenerate Conic

1. A plane intersects the vertex of the cone. _____
2. A plane intersecting one nappe, not at a right angle or parallel to the cone. _____
3. A plane intersecting both nappes. _____
4. A plane intersecting one nappe parallel to the side of the cone. _____
5. A plane intersecting a double cone at a right angle to the axis. _____

9-4 Circles

Match definitions and terms.

    A = Circle   B = Center   C = Radius   D = Chord   E = Diameter   F = Tangent

G = Secant   H = Interior   I = Exterior   J = Disk   K = Central Angle

L = Minor Arc   M = Major Arc   N = Circumference   O = Semicircle   P = Pi

1. Set of points with distance from the center less than the length of the radius. _____
2. Line that intersects a circle in exactly one point. _____
3. Angle whose vertex is at the center of the circle. _____
4. Arc that lies in the exterior of the central angle that intercepts the arc. _____
5. Distance from the center of the circle to a point on the circle. _____
6. Line that intersects a circle in two different points. _____
7. Set of points in a plane at an equal distance from a fixed point. _____
8. Value of the circumference divided by the diameter. _____
9. Chord that passes through the center of the circle. _____
10. Half of a circle created by drawing a diameter. _____
11. Set of points with distance from the center greater than the length of the radius. _____
12. Line segment whose end points lie on the circle. _____
13. Measurement around the outside of a circle. _____
14. Area enclosed by a circle. _____
15. Fixed point of the circle. _____
16. Arc that lies in the interior of the central angle that intercepts the arc. _____

Calculate the circumference, area, and arc length of a circle. The formulas are as follows:
$C = 2\pi r$. $A = \pi r^2$. Arc Length = $(\pi/180)$ x angle x radius. (Round to the nearest thousandth.)

17. Radius = 4, Central Angle = 60   _____   _____   _____
18. Radius = 6, Central Angle = 120   _____   _____   _____
19. Radius = 5, Central Angle = 45   _____   _____   _____
20. Radius = 3, Central Angle = 90   _____   _____   _____

Calculate the perimeter of a semicircle and area of a semicircle of a circle. The formulas are as follows: Perimeter of Semicircle = $r(\pi + 2)$. Area of Semicircle = $(\pi r^2)/2$. (Round to the nearest thousandth.)

21. Radius = 4   _____   _____
22. Radius = 6   _____   _____
23. Radius = 5   _____   _____
24. Radius = 3   _____   _____

9-5 Circular Sectors

Calculate the perimeter and area of a circular sector. The formulas are as follows: P = ((π/180) x angle x radius) + 2r. A = (angle x πr²) / 360. (Round to the nearest thousandth.)

1. Radius = 4, Central Angle = 60   _____ _____
2. Radius = 6, Central Angle = 120   _____ _____
3. Radius = 5, Central Angle = 45   _____ _____
4. Radius = 3, Central Angle = 90   _____ _____

9-6 Circular Segments

Calculate the perimeter and area of a circular segment. The formulas are as follows: P = ((π/180) x angle x radius) + c. A = ((angle x πr²) / 360) – (cd/2). D = 1/2√(4r²- c²). (Round to the nearest thousandth.)

1. Radius = 4, Central Angle = 60, Chord = 4   _____ _____
2. Radius = 6, Central Angle = 120, Chord = 10.25   _____ _____
3. Radius = 5, Central Angle = 45, Chord = 3.8125   _____ _____
4. Radius = 3, Central Angle = 90, Chord = 4.25   _____ _____

9-7 Circular Rings

Calculate the area of a circular ring. The formula is as follows: A = π (r1² – r2²). (Round to the nearest thousandth.)

1. Radius 1 = 5, Radius 2 = 4   _____
2. Radius 1 = 6, Radius 2 = 3   _____
3. Radius 1 = 8, Radius 2 = 7   _____
4. Radius 1 = 10, Radius 2 = 8   _____

9-8 Circular Ring Sectors

Calculate the area of a circular ring sector. The formula is as follows: A = ((angle x πr1²) / 360) – ((angle x πr2²) / 360). (Round to the nearest thousandth.)

1. Radius 1 = 5, Radius 2 = 4, Central Angle = 180   _____
2. Radius 1 = 6, Radius 2 = 3, Central Angle = 30   _____

3. Radius 1 = 8, Radius 2 = 7, Central Angle = 75 _____
4. Radius 1 = 10, Radius 2 = 8, Central Angle = 15 _____

• 9-9 Ellipses

Calculate the area and circumference of an ellipse. The formulas are as follows: A = πab. C = π(a + b). (Round to the nearest thousandth.)

1. Half Major axis = 4, Half Minor Axis = 3 _____ _____
2. Half Major axis = 4, Half Minor Axis = 3 _____ _____
3. Half Major axis = 4, Half Minor Axis = 3 _____ _____
4. Half Major axis = 4, Half Minor Axis = 3 _____ _____

• 9-10 Parabolas

Calculate the area of a parabola. The formula is as follows: A = 2/3(bh). (Round to the nearest thousandth.)

1. Base = 3, Height = 6 _____
2. Base = 4, Height = 5 _____
3. Base = 5, Height = 2 _____
4. Base = 6, Height = 4 _____

• 9-11 Hyperbolas

Match definitions and terms.

A = Foci   B = Transverse Axis   C = Center   D = Vertex
E = Conjugate Axis   F = Asymptote Line

1. The boundary of the hyperbola curve. ____
2. The point on the branch closest to the center. ____
3. A line passing through the foci. ____
4. A line perpendicular to the transverse axis and passes through the center. ____
5. The midpoint between the branches. ____
6. Two fixed points used to define a hyperbola. ____

9-12 Stadiums

Calculate the area and perimeter of a stadium. The formulas are as follows: $A = \pi r^2 + 2ra$. $P = 2(\pi r + a)$. (Round to the nearest thousandth.)

1. Radius = 1, Side Length = 4   _____   _____
2. Radius = 2, Side Length = 8   _____   _____
3. Radius = 3, Side Length = 6   _____   _____
4. Radius = 4, Side Length = 10   _____   _____

9-13 Ovals

Mark as True or False.

1. An oval is a shape consisting of a closed curve.   ____
2. An oval does not have a precise mathematical definition.   ____
3. The area of an oval uses the same formula as a stadium.   ____
4. All ovals have the exact same shape.   ____

CHAPTER 10 Three Dimensional Non-polytopes

10-1 Introduction

Three dimensional non-polytopes are geometric objects with three dimensions that do not have straight sides. The sides of non-polytopes are curved. This section will describe the following shapes: sphere, cone, cylinder, conical frustum, torus, spherical cap, spherical sector, spherical segment, spherical wedge, and capsule. The last five objects are derived from a sphere or cylinder.

10-2 Spheres

A sphere is the set of all points in three-dimensional space at an equal distance from a fixed point. This set of points forms a continuous closed surface. The fixed point is the center of the sphere. The distance from the center of the sphere to a point on the sphere is called the radius. A diameter is a chord that passes through the center of the sphere. A radius equals one half of a diameter.

The interior of a sphere is the set of all points whose distance from the center is less than the length of the radius of the sphere. The exterior of a sphere is the set of all points whose distance from the center is greater than the length of the radius of the sphere. The area enclosed by a sphere is called a ball. The area of a ball is commonly referred to as the area of a sphere.

Pairs of points on the sphere located on opposite ends of a diameter are called antipodes or antipodal points. A great circle is created by the intersection of a plane and a sphere, with the plane passing through the center of the sphere. It cuts the sphere into two equal parts or hemispheres. The intersection of a plane and a sphere, but not through the center, creates a small circle. The shortest path between two non-antipodal points on the surface of a sphere, known as the orthodome, lies on a unique great circle that passes through the two points. The arc of the great circle between the two points is called a geodesic. If the two points are antipodal, then there are an infinite number of shortest paths between them. The first two figures below show a sphere and a sphere with a great circle.

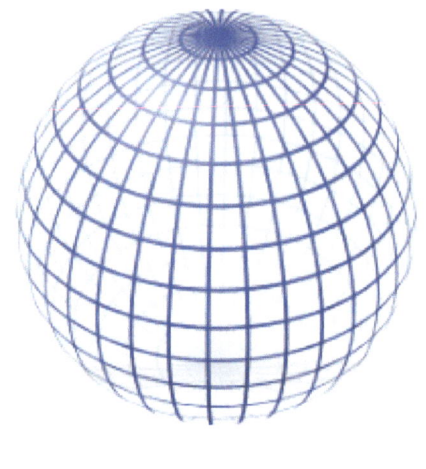
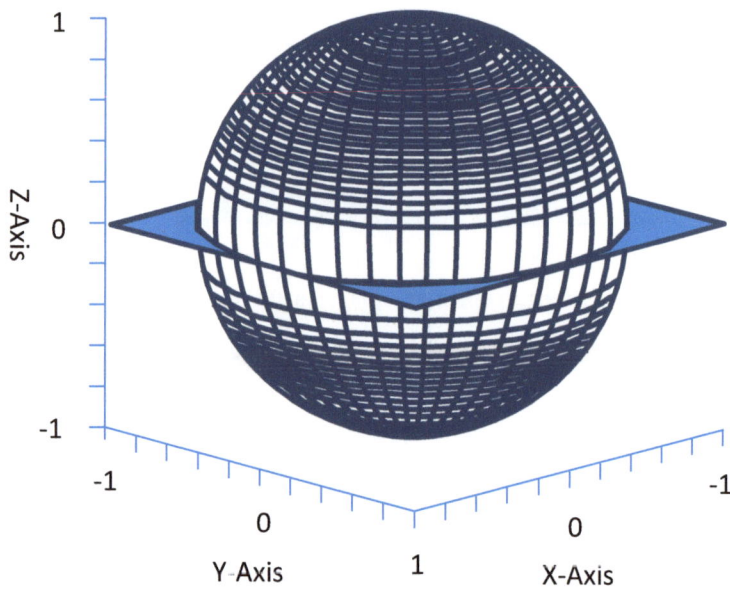

Figure 10-1: Sphere

Figure 10-2: Sphere – Great Circle

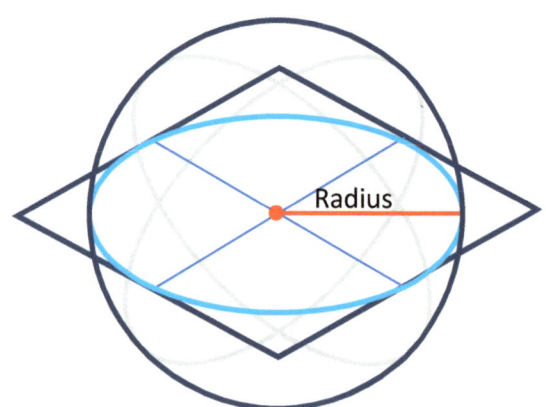
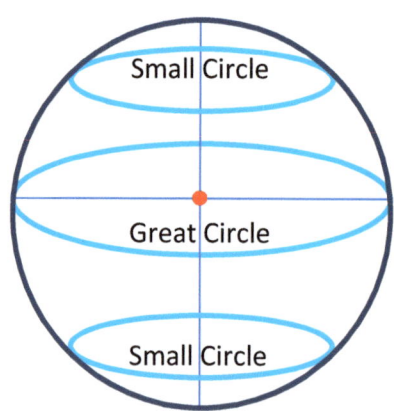

Figure 10-3: Sphere - Radius

Figure 10-4: Sphere – Great and Small Circles

The last two figures above show a sphere with a radius and a sphere with a great circle and small circles. The **circumference** is the length of any great circle, the intersection of the sphere with any plane passing through its center. A **meridian** is any great circle passing through a point designated a pole. A **geodesic**, the shortest distance between any two points on a sphere, is an arc of the great circle through the two points. Of all the shapes, a sphere has the smallest surface area for a volume. It can contain the greatest volume for a fixed surface area.

The surface area of a sphere can be written as: $S = 4\pi r^2$. The volume of a sphere can be written as: $V = 4/3\ \pi r^3$. If the radius of a sphere is 2 units, then the surface area is 50.2654 square units and the volume of the sphere is 33.5103 cubic units.

• 10-3 Spherical Caps

A spherical cap is the region of a sphere which lies above, or is cut off by, a plane. The surface area of a spherical cap can be written as: $S = 2\pi rh$ or $S = \pi(a^2 + h^2)$. The volume of a spherical cap can be written as: $V = 1/6\ \pi h(3a^2 + h^2)$ or $V = 1/3\ \pi h^2(3 - h)$. The radius of the base circle can be written as: $a = \sqrt{(h(2r - h))}$. The radius of the sphere can be written as: $r = (a^2 + h^2) / 2h$. The figure below shows a spherical cap. The second figure below shows a detailed drawing of a spherical cap segment with labeled parts.

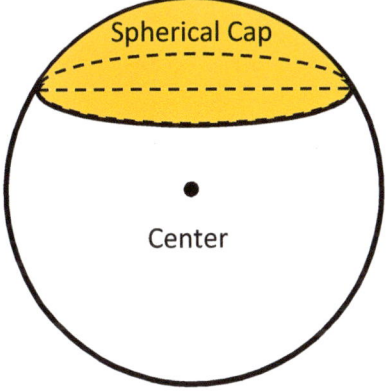

Figure 10-5: Spherical Cap

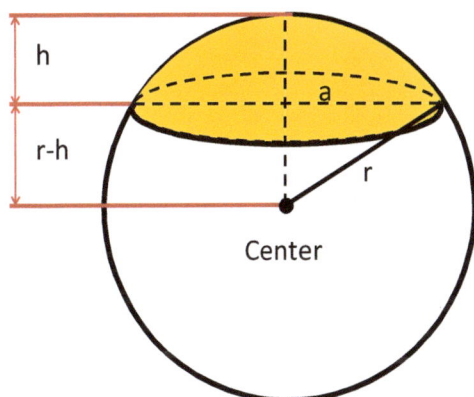

Figure 10-6: Spherical Cap - Detail

• 10-4 Spherical Sectors

A spherical sector is the region of a sphere bounded by two radii and their intercepted angle. It is a cone shaped piece of a sphere. The arc is based on the central angle formed by the two radii. The spherical sector may be open with a conical hole or closed as a spherical cone. The surface area of a spherical sector can be written as: $S = 2\pi rh + \pi ar + \pi br$. For an open spherical sector, the formula simplifies to: $S = \pi r(2h + a + b)$. For a spherical cone, where $b = 0$, the formula simplifies to: S = πr(2h + a). The volume of a spherical sector, either open or closed, can be written as: V = 2/3 πr²h. The figures below show an open spherical sector and a spherical cone.

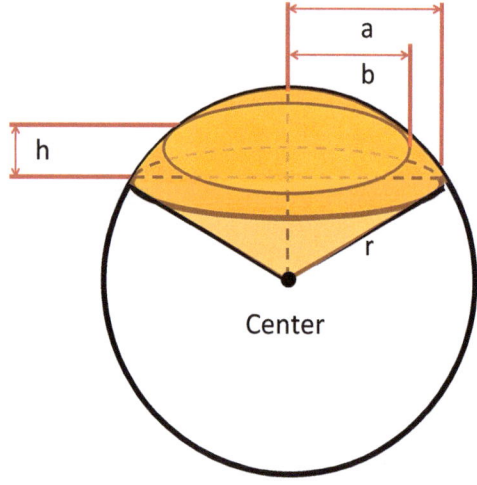

Figure 10-7: Open Spherical Sector

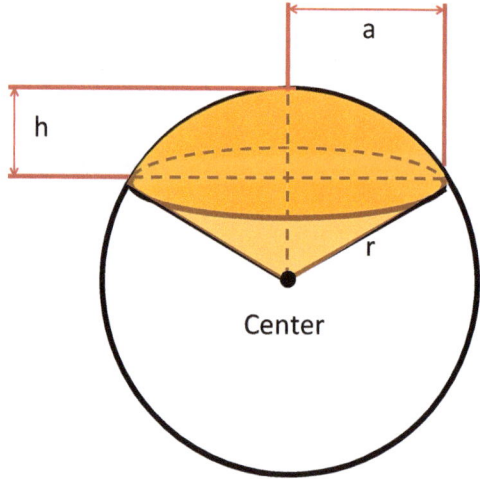

Figure 10-8: Spherical Cone

10-5 Spherical Segments

A spherical segment or spherical frustum is the region of a sphere which lies between two parallel planes. The surface of a spherical segment, excluding the two bases, is called a spherical zone. The surface area of a spherical zone can be written as: S = 2πrh. The surface area of the spherical segment, including the two bases, can be written as: S = 2πrh + πa² + πb². The volume of a spherical segment can be written as: V = 1/6 πh(3a² + 3b² + h²). The figure below shows a spherical segment.

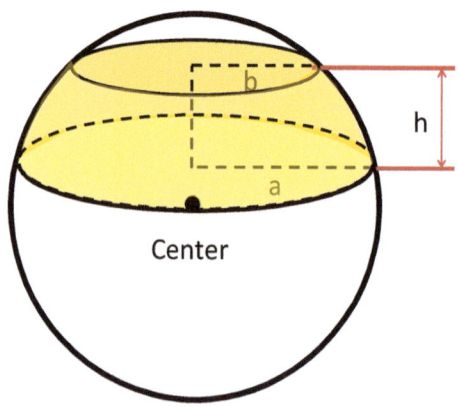

Figure 10-9: Spherical Segment

10-6 Spherical Wedges

A spherical wedge is a portion of a sphere bounded by two plane semidisks and their intercepted angle. It is formed by revolving a semicircle about its Z-axis diameter by less than 360 degrees. An angle of $180°$ produces a hemisphere and an angle of $360°$ produces a ball. The surface of a spherical wedge is called a spherical lune. The surface area of a spherical lune can be written as: $S = (4\pi r^2)$(angle in degrees/360) or $S = 1/90 \, (\pi r^2)$(angle in degrees). The volume of a spherical wedge can be written as: $V = (4/3 \, \pi r^3)$(angle in degrees/360) or $1/270 \, (\pi r^3)$(angle in degrees). The figures below show a spherical wedge and a spherical lune.

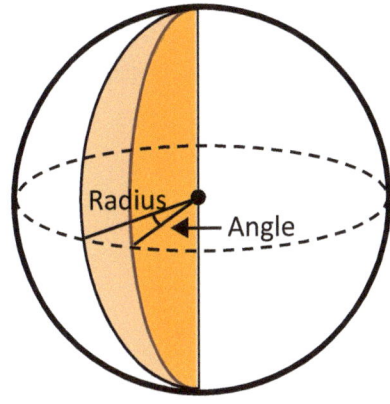

Figure 10-10: Spherical Wedge

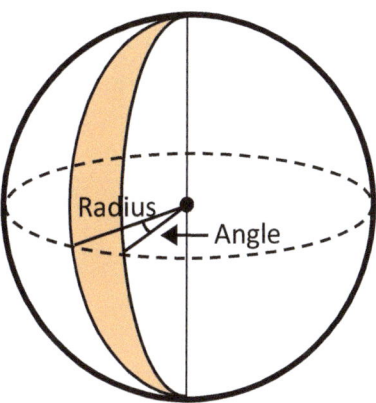

Figure 10-11: Spherical Lune

10-7 Cones

A cone is a non-polyhedron in the form of a conic solid. A conic solid is a three-dimensional geometric shape bounded by a plane base and the surface formed by line segments connecting the perimeter of the base to a common point, or apex, outside the plane of the base. A conic solid with a circular base is a cone. The base of a cone is a circle. The side of the cone, called a lateral surface, is always curved. The apex where all of the lines of the sides meet is called the vertex of the cone. The height of a cone is the perpendicular distance from the vertex to the base. The figures below show the parts of a cone.

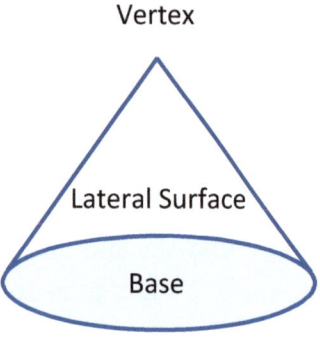

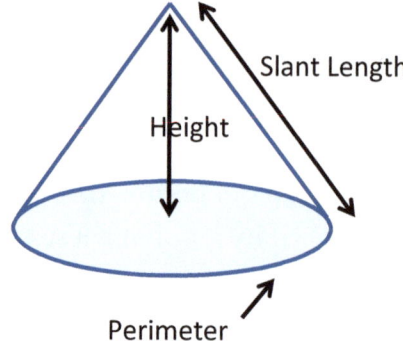

Figure 10-12: Cone – Parts

Figure 10-13: Cone - Measurements

Cones can be classified based on the location of the vertex. Cones can be right or oblique. If an axis through the vertex and the center of the base, the altitude, intersects the base perpendicularly, then the cone is a **right cone**. If the axis does not intersect perpendicularly, then the cone is an **oblique cone**. In a right cone, the height and altitude are the same line or axis. The figures below show right and oblique cones.

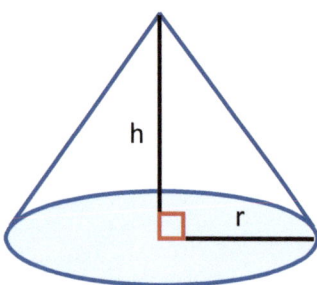

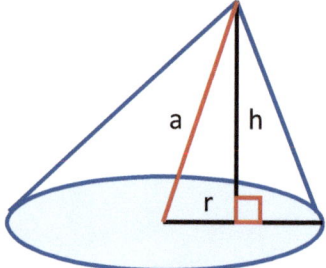

Figure 10-14: Right Cone

Figure 10-15: Oblique Cone

The surface area of a cone is the sum of the base area and the lateral surface area. The surface area of the base of a cone can be written as: S base = πr^2. The surface area of the lateral surface of a cone can be written as: S lateral = $\pi r s$, with s as the slant height. If the slant height is not known, then the surface area of the lateral surface can be written as: S lateral = $\pi r \sqrt{(r^2 + h^2)}$. The surface area of a cone can be written as: S = $\pi r^2 + \pi r s$ or S = $\pi r^2 + \pi r \sqrt{(r^2 + h^2)}$. The volume of a cone can be written as: V = 1/3 bh, with b the area of the base. If the area of the base is unknown, then the volume of a cone can be written as: V = 1/3 $\pi r^2 h$. The volume of a cone is equal to one third of the volume of a cylinder with the same base and height. The **center of mass** of a cone is located at a point on the axis one fourth of the height from the base.

10-8 Cylinders

A cylinder is constructed from two congruent circular bases located in parallel planes and the lateral surface connecting the two bases. All of the points of the lateral surface are an equal distance from the central axis of the cylinder. If the central axis intersects the bases perpendicularly, then the cylinder is a right cylinder. If the axis does not intersect perpendicularly, then the cylinder is an oblique cylinder. In a right cylinder, the height and altitude are the same line or axis. The figures below show right and oblique cylinders.

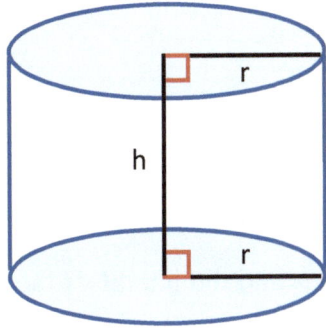

Figure 10-16: Right Cylinder

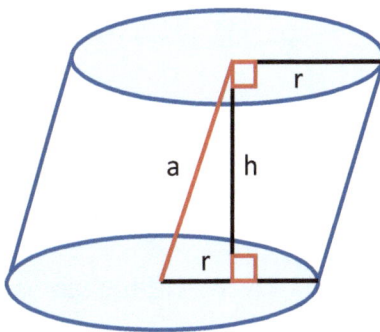

Figure 10-17: Oblique Cylinder

The surface area of a cylinder is the sum of two base areas and the lateral surface area. The surface area of the base of a cylinder can be written as: S base = πr^2. The surface area of the lateral surface of a cylinder can be written as: S lateral = $2\pi rh$. The surface area of a cylinder can be written as: S = $2\pi r^2 + 2\pi rh$ or S = $2\pi r(r+h)$. The volume of a cylinder can be written as: V = $\pi r^2 h$. The volume of a cylinder is equal to three times the volume of a cone with the same base and height. The center of mass of a cylinder is located at a point on the axis one half of the height from the lower base.

10-9 Conical Frusta

A conical frustum is a shape with properties common to both a cone and a cylinder. A conical frustum can be constructed by cutting off the top of a cone with a plane parallel to the base or by reducing the diameter of one base of a cylinder. The lateral surface slants from the base to a vertex as in a cone and the two bases are parallel as in a cylinder. A conical frustum can be right or oblique. The figures below show right and oblique conical frusta.

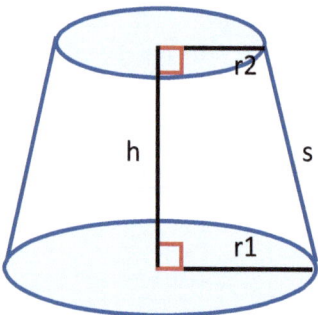

Figure 10-18: Right Conical Frustum

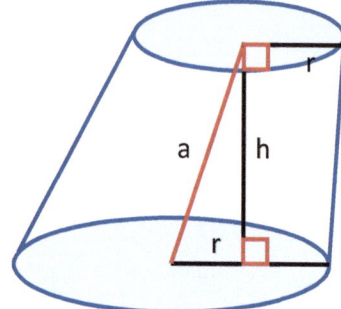

Figure 10-19: Oblique Conical Frustum

The surface area of a conical frustum is the sum of two base areas and the lateral surface area. The surface area of the bases of a conical frustum can be written as: S base1 = $\pi r1^2$ and S base2 = $\pi r2^2$ The surface area of the lateral surface of a conical frustum can be written as: S lateral = $\pi(r1 + r2)s$, with s as the slant height. If the slant height is not known, then the surface area of the lateral surface can be written as: S lateral = $\pi(r1 + r2)\sqrt{((r1 - r2)^2 + h^2)}$. The surface area of a conical frustum can be written as: S = $\pi(r1^2 + r2^2) + \pi(r1 + r2)s$ or S = $\pi(r1 + r2 + (r1r2)s)$. The volume of a conical frustum can be written as: V = $1/3\ \pi h(r1^2 + r1r2 + r2^2)$. The volume of a conical frustum can be calculated using the area of the two bases, a1 and a2, and can be written as: V = $1/3\ h(a1 + a2 + \sqrt{(a1a2)})$.

- ## 10-10 Tori

A torus is surface formed by revolving a circle in three-dimensional space around the z-axis. The surface is ring-shaped with a hole, similar to a three-dimensional circular ring. A torus can also be described as a cylinder that curves around to connect the two bases. The solid contained by the surface is called a toroid. The figures below show a torus and the circles of a torus.

Figure 10-20: Torus

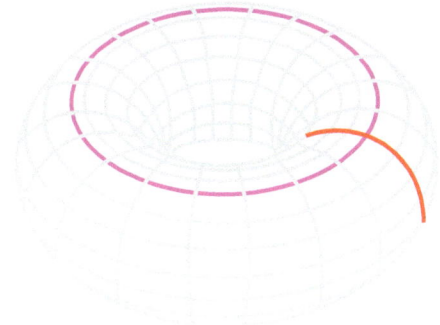

Figure 10-21: Circles of Torus

The figures below show two views of a torus with measurements, a side view and a top view.

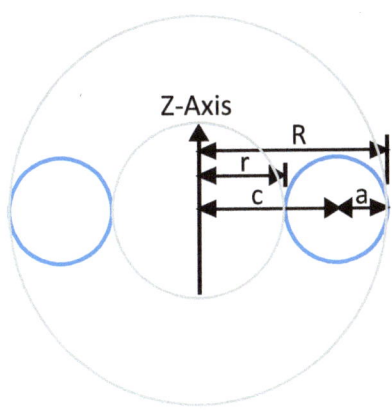

Figure 10-22: Torus – Side View

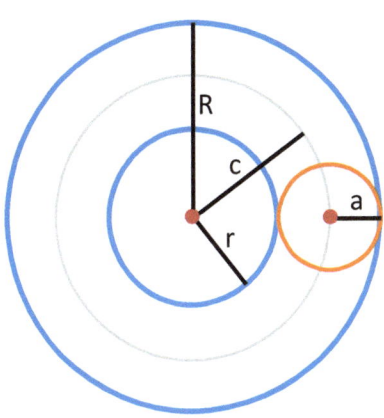

Figure 10-23: Torus – Top View

==Toroidal space== is the name used to describe the surface area and volume of a torus. A torus has ==four radius measurements==: inner radius, outer radius, average radius, and ring radius. The inner radius can be written as: $r = c - a$. The outer radius can be written as: $R = c + a$. The average radius can be written as: $c = 1/2 (R + r)$. The ring radius can be written as: $a = 1/2 (R - r)$. Using the radii of the torus, the surface area and volume can be calculated. The surface area of a torus can be written as: $S = (2\pi a)(2\pi c)$ or $S = 4\pi^2 ac$ or $S = \pi^2(R + r)(R - r)$. The volume of a torus can be written as: $V = (\pi a^2)(2\pi c)$ or $V = 2\pi^2 a^2 c$ or $V = 1/4\, \pi^2(R + r)(R - r)$.

10-11 Capsules

A capsule is a three-dimensional stadium or a cylinder with two hemispherical caps. The radius of the cylinder and the radius of the end cap are equal, and are called the radius of the capsule. The radius of the capsule can be written as: $r = c / 2\pi$. The circumference of the capsule can be written as: $2\pi r$. The length of the cylinder portion of the capsule can be written as: $a = (V/(\pi r^2)) - (4r/3)$ or $a = (S/2\pi r) - 2r$. The surface area of a capsule can be written as: $S = 2\pi r(2r + a)$. The volume of a capsule can be written as: $V = \pi r^2(4/3r + a)$. The figure below shows a capsule construction and capsule detail.

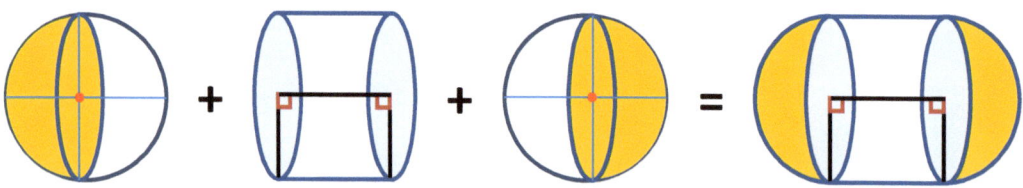

Spherical Cap + Right Cylinder + Spherical Cap = Capsule

Figure 10-24: Capsule Construction

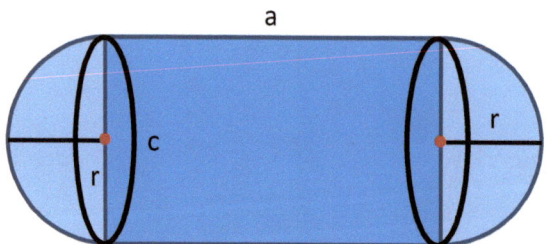

Figure 10-25: Capsule Detail

10-12 Summary

Three dimensional non-polytopes are geometric objects with three dimensions that do not have straight sides. The sides of non-polytopes are curved.

A sphere is the set of all points in three-dimensional space at an equal distance from a fixed point. This set of points forms a continuous closed surface. The fixed point is the center of the sphere. The distance from the center of the sphere to a point on the sphere is called the radius. A diameter is a chord that passes through the center of the sphere. A radius equals one half of a diameter.

A spherical cap is the region of a sphere which lies above, or is cut off by, a plane. A spherical sector is the region of a sphere bounded by two radii and their intercepted angle. It is a cone shaped piece of a sphere. A spherical segment or spherical frustum is the region of a sphere which lies between two parallel planes. The surface of a spherical segment, excluding the two bases, is called a spherical zone. A spherical wedge is a portion of a sphere bounded by two plane semidisks and their intercepted angle. It is formed by revolving a semicircle about its Z-axis diameter by less than 360 degrees. An angle of 180^0 produces a hemisphere and an angle of 360^0 produces a ball. The surface of a spherical wedge is called a spherical lune.

A cone is a non-polyhedron in the form of a conic solid. A conic solid is a three-dimensional geometric shape bounded by a plane base and the surface formed by line segments connecting the perimeter of the base to a common point, or apex, outside the plane of the base. A conic solid with a circular base is a cone. A cylinder is constructed from two congruent circular bases located in parallel planes and the lateral surface connecting the two bases. All of the points of the lateral surface are an equal distance from the central axis of the cylinder.

A conical frustum is a shape with properties common to both a cone and a cylinder. A conical frustum can be constructed by cutting off the top of a cone with a plane parallel to the base or by reducing the diameter of one base of a cylinder. The lateral surface slants from the base to a vertex as in a cone and the two bases are parallel as in a cylinder. A conical frustum can be right or oblique.

A torus is surface formed by revolving a circle in three-dimensional space around the z-axis. The surface is ring-shaped with a hole, similar to a three-dimensional circular ring. A torus can also be described as a cylinder that curves around to connect the two bases. The solid contained by the surface is called a toroid. A capsule is a three-dimensional stadium or a cylinder with two hemispherical caps. The radius of the cylinder and the radius of the end cap are equal, and are called the radius of the capsule.

CHAPTER 10

Chapter Test

Grading Scale: One point for each correct answer.

Excellent = 86-95, Good = 76-85, Average = 67-75, Fair = 57-66, Poor = 0-56

10-2 Spheres

Match definitions and terms.

A = Sphere   B = Center   C = Radius   D = Diameter   E = Interior   F = Exterior

G = Ball   H = Antipodal Points   I = Great Circle   J = Hemisphere

K = Small Circle   L = Orthodome   M = Geodesic   N = Circumference   O = Meridian

1. Distance from the center of the sphere to a point on the sphere. ____
2. Set of points with distance from the center greater than the length of the radius. ____
3. Any great circle passing through a point designated a pole. ____
4. Half of a sphere created by drawing a great circle. ____
5. Pairs of points on the sphere located on opposite ends of a diameter. ____
6. Set of points in three-dimensional space at an equal distance from a fixed point. ____
7. Length of any great circle and the measure around the outside of a sphere. ____
8. Shortest path between two non-antipodal points on the surface of a sphere. ____
9. Arc of the great circle between any two points. ____
10. Fixed point of the sphere. ____
11. Area enclosed by a sphere. ____
12. Intersection of a plane and a sphere, with the plane passing through the center. ____
13. Chord that passes through the center of the sphere. ____
14. Intersection of a plane and a sphere, but not through the center. ____
15. Set of points with distance from the center less than the length of the radius. ____

Calculate the surface area and volume of a sphere. The formulas are as follows: $S = 4\pi r^2$. $V = 4/3 \pi r^3$. (Round to the nearest thousandth.)

16. Radius = 3   _____   _____
17. Radius = 4   _____   _____
18. Radius = 5   _____   _____
19. Radius = 6   _____   _____

10-3 Spherical Caps

Calculate the surface area and volume of a spherical cap. The formulas are as follows: $S = 2\pi rh$. $V = 1/6\, \pi h(3a^2 + h^2)$. (Round to the nearest thousandth.)

1. Radius = 4, Height = 0.5, Chord = 4   _____ _____
2. Radius = 6, Height = 3.25, Chord = 10.25   _____ _____
3. Radius = 5, Height = 0.5, Chord = 3.8125   _____ _____
4. Radius = 3, Height = 0.75, Chord = 4.25   _____ _____

10-4 Spherical Sectors

Calculate the surface area of an open spherical sector and volume of a spherical sector. The formulas are as follows: $S = \pi r(2h + a + b)$. $V = 2/3\, \pi r^2 h$. A = Chord 1. B = Chord 2. (Round to the nearest thousandth.)

1. Radius = 4, Height = 0.5, A = 4, B = 3   _____ _____
2. Radius = 6, Height = 3.25, A = 10.25, B = 9.25   _____ _____
3. Radius = 5, Height = 0.5, A = 3.8125, B = 2   _____ _____
4. Radius = 3, Height = 0.75, A = 4.25, B = 3.5   _____ _____

10-5 Spherical Segments

Calculate the surface area and volume of a spherical segment. The formulas are as follows: $S = 2\pi rh + \pi a^2 + \pi b^2$. $V = 1/6\, \pi h(3a^2 + 3b^2 + h^2)$. R = Radius of sphere. H = Height of spherical zone. A = Radius of lower base. B = Radius of upper base. (Round to the nearest thousandth.)

1. Radius = 4, Height = 3.5, A = 4, B = 2   _____ _____
2. Radius = 6, Height = 4.5, A = 5, B = 3   _____ _____
3. Radius = 5, Height = 2.5, A = 3, B = 1   _____ _____
4. Radius = 3, Height = 1.5, A = 2, B = 1.5   _____ _____

10-6 Spherical Wedges

Calculate the surface area of a spherical lune and the volume of a spherical wedge. The formulas are as follows: $S = (4\pi r^2)(angle/360)$. $V = (4/3\, \pi r^3)(angle/360)$. (Round to the nearest thousandth.)

1. Radius = 4, Central Angle = 60   _____ _____
2. Radius = 6, Central Angle = 120   _____ _____

3. Radius = 5, Central Angle = 45 _____ _____
4. Radius = 3, Central Angle = 90 _____ _____

10-7 Cones

Calculate the surface area and volume of a cone. The formulas are as follows: $S = \pi r^2 + \pi r s$. $V = 1/3\, \pi r^2 h$. (Round to the nearest thousandth.)

1. Radius = 2, Height = 1, Slant Height = 1.5 _____ _____
2. Radius = 3, Height = 1.5, Slant Height = 2.5 _____ _____
3. Radius = 4, Height = 2, Slant Height = 3 _____ _____
4. Radius = 5, Height = 2.5, Slant Height = 4 _____ _____

10-8 Cylinders

Calculate the surface area and volume of a cylinder. The formulas are as follows: $S = 2\pi r(r+h)$. $V = \pi r^2 h$. (Round to the nearest thousandth.)

1. Radius = 2, Height = 7 _____ _____
2. Radius = 4, Height = 5 _____ _____
3. Radius = 6, Height = 3 _____ _____
4. Radius = 8, Height = 1 _____ _____

10-9 Conical Frusta

Calculate the surface area and volume of a conical frustum. The formulas are as follows: $S = \pi(r1 + r2 + (r1 r2)s)$. $V = 1/3\, h(a1 + a2 + \sqrt{(a1 a2)})$. S base1 (a1) = $\pi r1^2$ and S base2 (a2) = $\pi r2^2$. R1 = Radius of Base 1. R2 = Radius of Base 2. (Round to the nearest thousandth.)

1. R1 = 5, R2 = 2, Height = 2, Slant Height = 2.5 _____ _____
2. R1 = 7, R2 = 1, Height = 4, Slant Height = 5 _____ _____
3. R1 = 4.5, R2 = 3, Height = 1, Slant Height = 1.5 _____ _____
4. R1 = 8, R2 = 5, Height = 2, Slant Height = 2.5 _____ _____

10-10 Tori

Calculate the surface area and volume of a torus. The formulas are as follows: $S = (2\pi a)(2\pi c)$. $V = (\pi a^2)(2\pi c)$. C = Average radius. A = ring radius. (Round to the nearest thousandth.)

1. Average Radius = 3, Ring Radius = 1   _____   _____
2. Average Radius = 4, Ring Radius = 2   _____   _____
3. Average Radius = 5, Ring Radius = 3   _____   _____
4. Average Radius = 6, Ring Radius = 4   _____   _____

10-11 Capsules

Calculate the surface area and volume of a capsule. The formulas are as follows: $S = 2\pi r(2r + a)$. $V = \pi r^2(4/3r + a)$. (Round to the nearest thousandth.)

1. Radius = 1, Side Length = 4   _____   _____
2. Radius = 2, Side Length = 8   _____   _____
3. Radius = 3, Side Length = 6   _____   _____
4. Radius = 4, Side Length = 10   _____   _____

www.ingramcontent.com/pod-product-compliance
Lightning Source LLC
Chambersburg PA
CBHW051025180526
45172CB00002B/473